W9-BDE-982

The McGraw-Hill 36-Hour Business Statistics Course

Other Books in The McGraw-Hill 36-Hour Course Series

Bittel
The McGraw-Hill 36-Hour Management Course

Dixon
The McGraw-Hill 36-Hour Accounting Course

Seglin
The McGraw-Hill 36-Hour Marketing Course

Schoenfield and Schoenfield
The McGraw-Hill 36-Hour Negotiating Course

The McGraw-Hill 36-Hour Business Statistics Course

Robert Rosenfeld
Professor of Mathematics and Statistics
Nassau Community College

McGraw-Hill, Inc.
New York St. Louis San Francisco Auckland Bogotá
Caracas Lisbon London Madrid Mexico Milan
Montreal New Delhi Paris San Juan São Paulo
Singapore Sydney Tokyo Toronto

Library of Congress Cataloging-in-Publication Data

Rosenfeld, Robert.
The McGraw-Hill 36-hour business statistics course / Robert Rosenfeld.
p. cm.
Includes bibliographical references and index.
ISBN 0-07-053837-9 (hc) : —ISBN 0-07-053836-0 (pb)
1. Management—Statistical methods. 2. Statistics. I. McGraw-Hill, Inc. II. Title. III. Title: McGraw-Hill thirty-six-hour business statistics course.
HD30.215.R67 1992
519.5'02465—dc20 92-4581
CIP

1 2 3 4 5 6 7 8 9 0 DOC/DOC 9 8 7 6 5 4 3 2

ISBN 0-07-053837-9 {HC}
ISBN 0-07-053836-0 {PBK}

The sponsoring editor for this book was James H. Bessent, Jr., and the production supervisor was Donald F. Schmidt. It was composed in Baskerville by North Market Street Graphics.

Printed and bound by R. R. Donnelley & Sons Company.

Passat is a registered trademark of Volkswagen Corp.

Contents

Preface

This book is designed for self-study. As such, the focus is on reader familiarization with the basic vocabulary of statistics. The goal is the promotion of statistical literacy, meaning the ability to follow a discussion which involves statistics and to think critically about conclusions which have been made on the basis of statistics. Different from a standard classroom text, it has relatively few exercises, and does not stress technical subtleties. It is not intended to teach the reader to become a professional statistician, but to come to appreciate the issues that are of concern.

The choice of material is based on over 25 years of personal experience as a statistical consultant and teacher for diverse audiences, which include experienced business and medical professionals, as well as beginning students in these fields. I have chosen to stress those ideas which are frequently used, but which are easily misunderstood. The first part of the book presents standard methods of organizing, summarizing, displaying, and interpreting numerical information. Then follows the basic concepts of probability and the key methods, which are based on probability, for making reasonable inferences about large groups of people based on data collected from representative smaller subgroups. Some of the specific topics covered include the use of averages, percentages, rates, and indexes; the principles of estimation, especially the assignment of a margin of error; the description of trends; and some standard approaches for evaluating statistical evidence.

Robert Rosenfeld

Acknowledgments

I would like to thank John Snell, Karl Bissex, Edith Schubert, and Dean Shattuck for providing material or suggestions for the book. For encouragement and comfort in times of stress, thanks to my wife, Leda, and, of course, our dog, Smee. For his help in getting this book started, and as one more tribute to his uncompromising integrity as a businessman, this book is dedicated to the memory of my friend, Michael Cyprian.

The McGraw-Hill 36-Hour Business Statistics Course

Introduction

This book, which is an introduction to the ideas of basic statistics that are most useful to managers, was designed to be brief rather than long, concrete rather than abstract, and friendly rather than intimidating. Its purpose is to point out how statistics, by which is meant the collection, presentation, and interpretation of numerical information, can be a vital aid to clear thinking in business. Statistics is not a god to be worshipped, or a substitute for your own common sense and experience, but it can help you to clarify points you wish to make, and to analyze the presentations of others. It can be the difference between having only a rough idea about the state of affairs and having precise insight. Perhaps, more valuable than anything else, it can help you think critically about your own intuitive impressions.

This short course stresses the fundamentals necessary to understand and interpret statistics as commonly experienced in business settings. It presents basic ideas and points of view intended to make dealing with statistics a sensible and nonthreatening affair. It is designed to make you a more intelligent user of statistics, not a professional statistician. A bibliography is included at the end of the book as a guide to more advanced resources.

Program of Study

The material is best covered in the order written. It is divided into nine chapters, each of which will take about four hours of study. Completing one chapter per week for nine consecutive weeks is a good pace for completing the material.

As in all study that involves mathematics, you learn best by *doing* problems. A good part of the time devoted to a chapter, therefore, will be used for working out examples and exercises. It is not usually until you *try* to solve a problem that you find out if you really understand how to do it. Because you will be doing writing and calculating, you will find it helpful to set aside a notebook in which to record results, and to use a calculator to save time with computation.

Here is one suggested approach for efficient study of a chapter.

1. First quickly browse through the chapter, noting the key concepts which are listed in a group at the start of the chapter and also noting the section headings.
2. Go slowly through the chapter, using a notebook to check all the calculations and interpretations in the examples.
3. After you have studied an example in the chapter, then write a brief comment in your notes which explains what the *point* of the example is. If it is appropriate, make up a similar scenario more precisely tuned to your own work.
4. After you have completed the chapter, reread the list of key concepts to check if all are clear.
5. Check the vocabulary list ("The Language of Statistics") which is part of every chapter, especially noting any expressions which are new to you. Write a sentence or two using each item in the list so that you feel comfortable with the word or phrase.
6. As a final check of your understanding, solve the problems at the end of the chapter. The answers are given in an appendix at the back of the book.

The Final Examination

To provide evidence of successful completion of this course, the optional "Final Examination" appears at the end of the book. This examination consists of 60 multiple-choice questions. You may take the examination and send it to the McGraw-Hill certification examiner for grading. A score of 70 percent or better entitles you to a certificate of accomplishment presented by McGraw-Hill, Inc. Details are provided with the examination at the end of the book.

1
Overview

Key Concepts

1. Statistics is an aid to decision making, but not a substitute for good business sense.
2. Descriptive statistics is the use of numerical information to summarize, simplify, and present masses of data.
3. Inferential statistics is the use of numerical information to make decisions or estimates about populations on the basis of limited information taken from samples of those populations.

The Language of Statistics

Inferential statistics

Descriptive statistics

Population

Sample

Margin of sampling error

What Is Statistics?

Key Concept 1

Statistics is an aid to decision making, but not a substitute for good business sense.

It's not surprising that so many people are leery of statistics. It has a reputation as a tool of charlatans and obfuscators, or more recently, the "number crunchers." The old saw that you can prove anything with statistics is enough to discourage anyone from making a serious attempt to learn to use statistics well. The main focus of this book is to present a few basic ideas which can help a manager use statistics intelligently and to interpret the statistical presentations of others sensibly. Above all, I'll stress the use of statistics as an aid to clear thinking. Statistics is an important tool to help a business manager organize numerical information, present it clearly, and make decisions, but it is not a substitute for good business sense and experience. A recent headline in the *New York Times* illustrates the point. It said, "Whatever the statistics show, is a subway ride worth $1.11 or $1.15 or $1.25?" The statistics can show the profit and loss figures for the various fare schedules, but anticipating what the riders of New York will find acceptable takes business and political sense in addition to statistics.

Coming to Terms with the Several Meanings of Statistics

A simple definition of statistics is hard to give because it is a word that has several overlapping meanings. It can refer simply to collections of numerical data. When you look at last year's sales figures, you are looking at statistics. But it may also refer to the summary measures calculated from these figures. In this sense, the average sales per month is a statistic computed from the 12 monthly sales numbers.

In a somewhat different sense, statistics refers specifically to the *activity* of using and interpreting the information. In this sense, a manager *does* statistics. An example of this is making projections, such as estimating next quarter's sales.

In general, though, we expect statistics to deal one way or another with large quantities of numbers. The word itself is derived from "state," and its earliest use applied to the study of political facts and figures, and to the keeping and analysis of tables of political information.

Descriptive Statistics

Key Concept 2

Descriptive statistics is the use of numerical information to summarize, simplify, and present masses of data.

The use of statistics in business can be broadly characterized as either descriptive or inferential. **Descriptive** is what is meant when data is organized and summarized for clearer presentation. Most projects begin with the accumulation of information, which is difficult to interpret until it is reorganized in some sensible way. This organization often includes arranging the data into tables, or representing the data graphically. Further, it may be useful to summarize and simplify the data by reporting numerical measures like averages or percentages. A clear summary of last year's sales or of current payroll expenses is a familiar example of descriptive statistics. For ease of communication it is often better to start with neat summaries, and only go to the mass of raw data when it is necessary to clarify some detail.

Example 1-1: A random sample of 200 shoppers agreed to give their opinions on a new product. Table 1-1 gives a general summary of their responses, and Figure 1-1 presents two simple graphical versions of this information, a bar graph and a pie chart.

Table 1-1 and Figure 1-1 are fairly crude summaries which describe the overall reception of the product. They make a good starting point. They

Table 1-1. Opinions of 200 Shoppers about New Product

Opinion	Percentage
Liked product enough to buy it	54
Liked product but probably wouldn't buy it	32
Did not like product	14

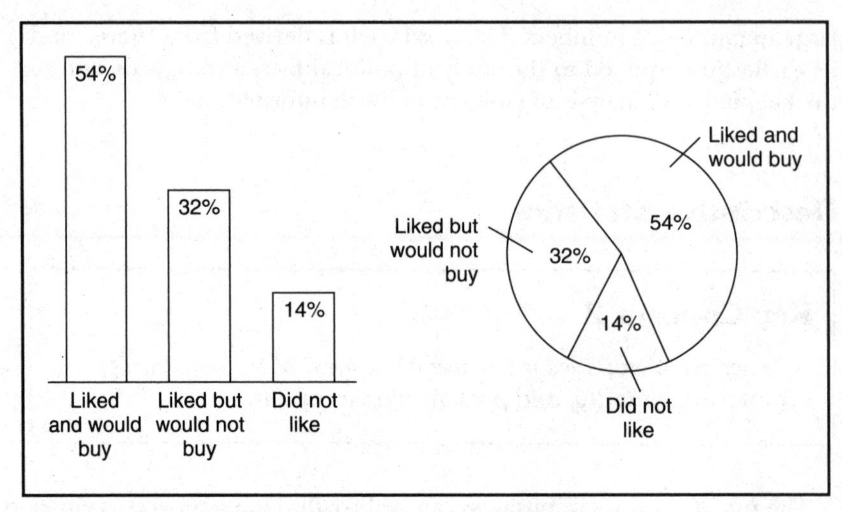

Figure 1-1. Bar graph and pie chart. Opinions of 200 shoppers about new product.

clearly show that the product is reasonably popular. But questions come to mind which call for more detail. For one, we might want to know more about the people who took part in the survey. This could be shown as in Table 1-2.

For a visual display of such information we might choose a bar graph, like the one in Figure 1-2.

With this level of detail you can see that the product was most popular with the shoppers in the 40 to 49 age group, and that the youngest and the oldest groups did not particularly like it. At this stage you might want to know other information about the shoppers in the sample, perhaps their sex or their incomes. As detail is called for, you can provide it in

Table 1-2. Description of the 200 Shoppers Interviewed, by Age Group

Age	Number that age in the sample		Number who liked the product enough to buy it	
15–19	10	(5% of sample)	1	(10% of age group)
20–29	20	(10%)	4	(20%)
30–39	72	(36%)	38	(52.7%)
40–49	76	(38%)	62	(81.6%)
50 and over	22	(11%)	3	(13.6%)
Total	200	(100%)	108	(54%)

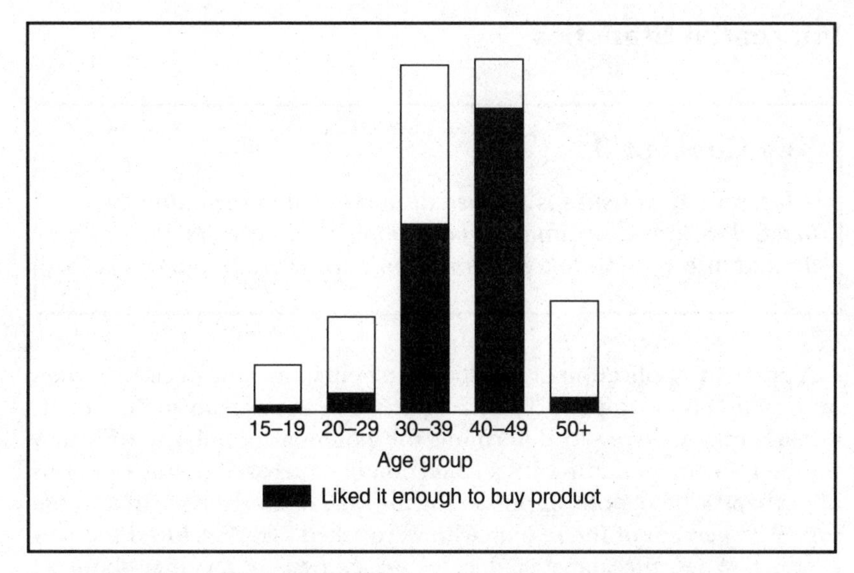

Figure 1-2. Bar graph. Response by age group of 200 shoppers to new product.

separate tables and graphs. If the main point for making the table or graph is to tell *other* people about the survey (as opposed to concise record keeping for yourself), it is a good idea to not include too much information in any one table or graph. The most extreme level of detail would be to list the opinions of all 200 people together with a description of each person. Except for the person responsible for analyzing the data, probably no one needs to know the results so thoroughly.

Note especially in Table 1-2 that both numerical counts and percentages were reported. Percentages without numerical counts make it difficult for the reader to evaluate the significance of the figures. If you only report percentages, then you should also give the total number of people in the survey. Percentages based on small numbers can be very misleading. When 1 person in a 2-person department is assigned to a new project you can certainly say that 50 percent of the department is devoted to the project, but it may not have the same impact as devoting 5 out 10 people or 20 out of 40, both of which are also 50 percent assignments.

I have used Example 1-1 to introduce the idea of descriptive statistics, the organization and presentation of numerical information collected in some study. The main point has been that the level of detail should be tailored to the intended audience. Chapters 2 and 3 describe in greater detail the technical components of descriptive statistics.

Inferential Statistics

Key Concept 3

Inferential statistics is the use of numerical information to make decisions or estimates about populations on the basis of limited information taken from samples of those populations.

A powerful application of statistics is to help in making decisions when information is incomplete. For example, when a sales manager conducts a marketplace survey to determine the potential popularity of a new product, she may end up with a collection of data based on the opinions of perhaps 1000 people. Suppose, for the sake of illustration, that in the survey 23 percent of the people who were asked said they loved the new product. When the manager decides *on the basis of this* that about 23 percent of *all* the *potential* buyers would love the product, she is making an *inference* from the sample. Any kind of generalization or prediction about how a larger group will behave based on information taken from only a smaller part of that group is called an **inference**. Statistically speaking, the large group is called the **population** and the smaller subgroup is called the **sample**. It is not necessary that these groups consist of *people;* in the technical language of statistics, the population is the large group of *values* from which the smaller group is taken. For example, the daily receipts for the past 365 days could be considered a population and the receipts for 50 of those days picked at random could be the sample.

There will be more about how to choose samples from populations in Chapter 5, but one critical point is that the sample included in the survey must be taken from the relevant population. This means two important things:

1. The population must be clearly identified; and
2. The rule for picking the sample must be clearly defined.

Decisions based on the results of faulty sampling are quite likely to be useless at best and disastrously expensive at worst. Furthermore, since inferences are always based on just a sample from the population, there is always some potential error in the inference. It is important to know the magnitude of this potential error.

Margin of Sampling Error

A sample of 90 hospitals across the country may indicate that 9 of them (10 percent) intend to buy a particular type of product in the next year. But we would not expect to find, one year later, that *precisely* 10 percent of *all* the hospitals in the country (there are over 6000) actually bought the product. The reasonable inference from the sample is that we expect *about* 10 percent of the hospitals to make the purchase. A good statistical report would specify the meaning of "about 10 percent."

Example 1-2: Suppose that as a result of a survey of 100 hospitals across the United States, we estimate that about 22 percent of all hospitals will buy product X during the next year. Suppose, further, that on the basis of statistical reasoning we determine that the **margin of error** on this estimate is 5 percent. This means that though the actual percentage may not turn out to be exactly 22 percent, we are quite confident it will be no less than 17 percent and no more than 27 percent. (These are the figures you get by adding and subtracting the margin of error to and from 22 percent.) This interval provides bounds to help estimate revenue from sales of the product.

When you need to evaluate a statistical claim that is obviously an inference from a sample, you should satisfy yourself that the sample is a legitimate representation of the appropriate population. And you should ascertain the margin of error. Without these two pieces of information, your ability to evaluate a claim is severely limited. The rules of statistics tell how to compute a margin of error, and experience in running surveys gives you insight into selecting samples. I will discuss both of these challenges in more detail later in the book, so for now it is adequate to just point out their importance.

Case Study

Statistics and Accounting

This is an example of a sampling procedure that was evaluated by comparing its results to the results achieved by calculations using *all* the data in the population. Professional statisticians designed the procedure, so it was not just a shot in the dark, but you might find the accuracy surprising.

The details are given by John Neter in the text *Statistics: A Guide to the Unknown*. He describes a six-month study by the Chesapeake and Ohio Railroad which showed that instead of a time-consuming enumeration of about 20,000 waybills (a document describing each freight transaction), a cleverly designed sampling procedure of about 2000 waybills projected virtually the same revenue for the railroad. The exact figures were:

Complete population: 22,984 waybills;
exact revenue due Chesapeake and Ohio: $64,651
Sample: 2072 waybills;
estimate of revenue due Chesapeake and Ohio: $64,568

In this case, a sample analysis which took much less time and cost much less money than a complete examination of all the data resulted in an estimate of revenue that was practically the same. The amount of the error was much less than the cost of doing a complete analysis. It might be pointed out that, in this case, the population is the collection of all 22,984 waybills, and the sample is the 2072 waybills actually used to estimate revenue. This study is an example of auditing by random sampling. This is a specialty of accountants trained in statistics.

One point to be taken from this case is that before you use a sampling procedure for business purposes, it would be advisable to carry out a pilot study which compared sample results to a complete evaluation of all the data, or to make a thorough investigation of the results of a similar project. Regardless of how clever the statistical calculation, the manager has to know that the sampling system works. And it would be a good idea to revalidate the process at appropriate intervals.

Self-Check

The answers to the Self-Check exercises can be found in Appendix A.

1. A manager prepares a report which shows monthly sales for the past year. This is an example of
   a. descriptive statistics
   b. inferential statistics.

For questions 2 to 5, read the paragraph and answer the questions.

During a market survey for a department store, information was collected by telephoning 300 customers picked at random from all those who held credit cards for the company. Among the questions on the survey was this one: Have you bought something from the store during the past six months?

2. Ten percent of the respondents answered "yes" to this question. If we conclude that about ten percent of *all* their credit card holders have made a purchase during the past six months then this is
   a. descriptive statistics
   b. inferential statistics

3. In this study the 300 people interviewed constitute
   a. the sample
   b. the population

4. If you want to know how accurate the ten percent estimate is, then you are asking for
   a. margin of error
   b. margin of inference

5. If the margin of error for the percentage of credit card holders who have made a purchase in the past six months is 3 percent, then what are the upper and lower bounds for the estimate?

Answer the following two questions.

6. In the case study on page 15 a sample about one-tenth the size of the population was used to estimate total revenue. Is this an example of inferential or descriptive statistics?

7. A car dealer says that his sales increased 100 percent this month compared to last month. Does this mean that he sold a lot more cars? Does it mean he made a lot more money?

2
Organizing and Displaying Data

Key Concepts

4. Two broad types of data are numerical (quantitative) and categorical (qualitative).
5. The main purpose of a graph is communication. The level of complexity must be appropriate for the intended audience.
6. Some useful and popular ways to display sets of data.

   Stem and leaf displays

   Frequency tables

   Histograms

   Bar graphs

   Pie charts

   Data maps

   Time series graphs
7. Be aware of the effect of changing the scale of a graph.

The Language of Statistics

Numerical data

Categorical data

Stem and leaf display

Frequency table

Interval width

Histogram

Pie chart

Bar graph

Time series graph

Types of Data: Numerical and Categorical

Key Concept 4

Two broad types of data are numerical (quantitative) and categorical (qualitative).

The responses to a survey, the observations in a study, and the measurements from an experiment comprise the data of a statistical report. Often these data are first collected and recorded in a fashion which makes immediate interpretation difficult. At some point data must be organized, summarized, and presented in tables or graphs. Only then do patterns appear which give them sense.

Statisticians classify data into two major types: numerical and categorical. **Numerical** (or quantitative) **data** consists of all data measured on a numerical scale. This includes information like sales figures, time expenditures, and number of employees—anything naturally recorded as a number. **Categorical** (or qualitative) **data** consists of data put into discrete classes, which often have no inherent numerical value, such as sales districts, departments in an organization, and brand names. We treat the two types of data a little differently because some calculations that make sense for one don't make sense for the other. For example, categorical data is usually presented by showing the counts or percentages in each of the classes, either in tables or in charts, while numerical data is presented by showing averages or other more complicated statistics or graphs.

Any characteristic of a population which is recorded is called a **variable.** If it is recorded by category, it is called a categorical, or qualitative, variable. If it is recorded as numerical data, it is called a quantitative variable.

Example 2-1: In a market survey each person is asked what brand of television set he or she bought most recently. The response is a name which is one of a set of discrete labels or categories. In this case, the variable, "Brand of TV," is a categorical or qualitative variable.

Example 2-2: The medical department of a corporation records the cholesterol level of any employee who uses the department for a physical exam. The response is a number which could be any value on a scale ranging from below 100 up to several hundred. This number is a measurement of the quantity of cholesterol in the person's blood. "Cholesterol level," then, is a numerical or quantitative variable.

Guidelines for Visual Displays

Graphs Must Make Sense

Let's discuss a few ways of organizing and displaying data, keeping in mind that technology, especially computer graphics, makes it easier every year to produce elaborate and even spectacular displays. The point to keep in mind always is communication—does the picture make sense? A good graph or table makes it easy to see what's going on. A simple and uncluttered picture can be more effective than a crowded or poorly designed one. One author who has valuable detailed suggestions for graphs that work well is Edward R. Tufte of Yale University; see the bibliography for titles of his books.

Example 2-3: To underscore the point made in the previous paragraph, look at Figure 2-1, which is typical (though somewhat simplified) of graphs shown in computer magazines. In some ways this is a most attractive graph, especially when printed in color, but for certain information it is not easy to read. Try, for example, to compare Brand B and Brand F in the category "Speed." In response to graphs like this one, one reader wrote to the editor of a popular PC monthly, "Your 3-D bar charts look great. They have tremendous impact, and with that classy shading they really jazz your pages. Unfortunately, they are virtually impossible to interpret."

Have Presentation Graphs Previewed

Key Concept 5

The main purpose of a graph is communication. The level of complexity must be appropriate for the intended audience.

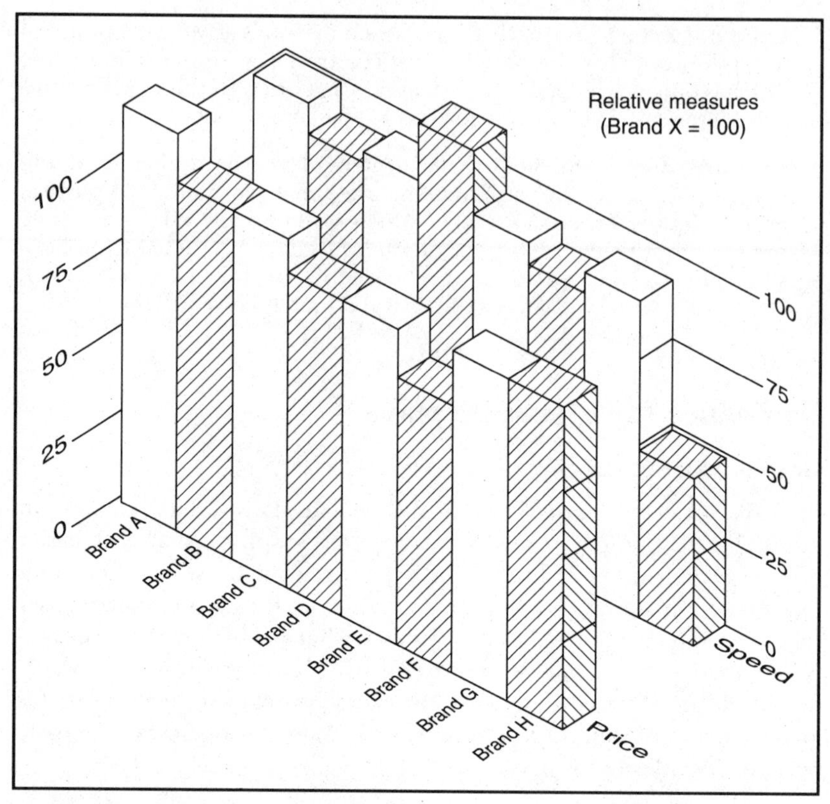

Figure 2-1. Three-dimensional multiple bar graph.

When you design a graph or table for presentation, it's worth the time spent to have it previewed by someone from the prospective audience. Let *them* tell you what they think it means; and then see if they got the right impression. This is particularly true for slides and overhead transparencies shown to large groups of people who cannot see small details, and who will not have very long to look at your picture. Keep it simple; do not put too much information on one graph. You can and should relax this rule for graphs and tables which people will have plenty of time to study where more complexity may be rewarding.

It is easy to forget that your data is clear to you because you have been immersed in it for weeks or months, while your audience may be seeing it for the first time. Anticipate questions generated by your displays and provide enough information so that a serious and interested viewer can figure out the answers to most of them.

Choose an Appropriate Graph

How you choose to display the data depends partly on the nature of the observations themselves. In one study you may simply be counting individuals in different categories: say, employees divided by sex or ethnic background. In another you may be taking measurements with great precision along some scale, such as the point at which a product breaks under stress, or the temperature at which a machine overheats. Graphs which are appropriate for one kind of data may not be appropriate for another. But rather than have a list of hard and fast rules, just keep in mind that the main point should be clarity of communication.

Displaying Numerical Variables

Key Concept 6

Some useful and popular ways to display sets of data.

Stem and leaf displays
Frequency tables
Histograms
Bar graphs
Pie charts
Data maps
Time series graphs

Consider a study designed to describe the time it takes to complete some task, say the duration of an initial phone call to a customer service department. Suppose you want to get an overall picture of the distribution of these times. You might begin by timing each call for some period of time. Because time is a continuous quantity, you have to decide how precisely to measure each call. With electronic devices, you could no doubt measure each call to the nearest hundredth of a second, but such precision is probably pointless. After giving it some thought, let's suppose you decide to measure 200 calls to the nearest second.

The raw data is shown in Table 2-1. Note that the table has a title as well as a table number. On the day that you create a table, you are cer-

Table 2-1. Lengths (in Seconds) of 200 Telephone Calls to the Customer Service Department

280	221	122	199	72	208	171	146	171	155
241	223	65	104	120	177	67	30	86	62
233	263	233	299	163	213	216	110	89	81
246	295	265	191	138	69	69	263	277	178
178	112	47	253	54	89	133	220	65	134
41	285	84	260	152	107	248	293	262	146
260	173	300	242	119	57	111	211	271	137
187	174	224	42	130	68	128	144	196	285
278	294	74	120	225	112	278	261	230	164
39	130	148	251	141	158	45	242	56	85
108	60	106	67	64	148	64	90	92	110
149	135	135	80	116	157	131	74	122	127
163	131	147	104	135	69	93	63	106	76
168	93	73	141	67	100	160	138	134	178
149	153	176	101	158	83	165	84	106	125
111	132	129	104	101	125	122	109	95	113
106	132	108	107	98	147	142	136	132	138
121	122	132	114	148	124	110	122	113	119
129	129	112	121	118	111	116	127	128	110
115	121	123	119	128	123	123	127	118	121

tain that you will never forget what the numbers represent. But sooner or later you will, and then the title becomes invaluable. And certainly other readers or interpreters will find the title helpful.

How can you display all this information so that it conveys a sense of what is going on? Below, several ways are suggested that are not too difficult to do by hand, and are also easily produced by computer.

Stem and Leaf Display

The **stem and leaf display** in Figure 2-2 can be read as follows. In the top row, the 3 indicates that all the numbers in that row are in the "thirties." In particular there is a 30 (indicated by the 0) and a 39 (indicated by the 9). In the next row, the 4 indicates that the numbers are in the "forties," and so forth. The values to the left of the vertical line are the "stems"; on the right side are the "leaves." Because the stems jump in increments of 10 seconds, we say that the unit for the stems is 10; similarly, the units for the leaves is 1.

Because the stem and leaf display orders the observations it is easy to see that the shortest phone call was 30 seconds and that the longest was 300 seconds. Further we see some "shape" to the data. It can be seen that there are quite a few calls between 100 seconds and 150 seconds. The long rows of numerals create a bulge in the display. If we rotate the display and make a smooth curve by tracing its outline as in Figure 2-3 we can also see that the shape has a long "tail" on the right side. Statisticians describe this by saying that the distribution of observations is mound-shaped with a *heavy tail* on the right side.

Frequency Tables

For the phone call data in Table 2-1, even the stem and leaf diagram may be too bulky for easy comprehension because the range of values is so great. It would be convenient to collapse the data into wider intervals. This can be done with a **frequency table**.

Table 2-2 presents a frequency table using intervals of *width* 20 seconds. This means that each interval starts 20 units after the previous one. The first interval has been arbitrarily started at 30 seconds. (It could just

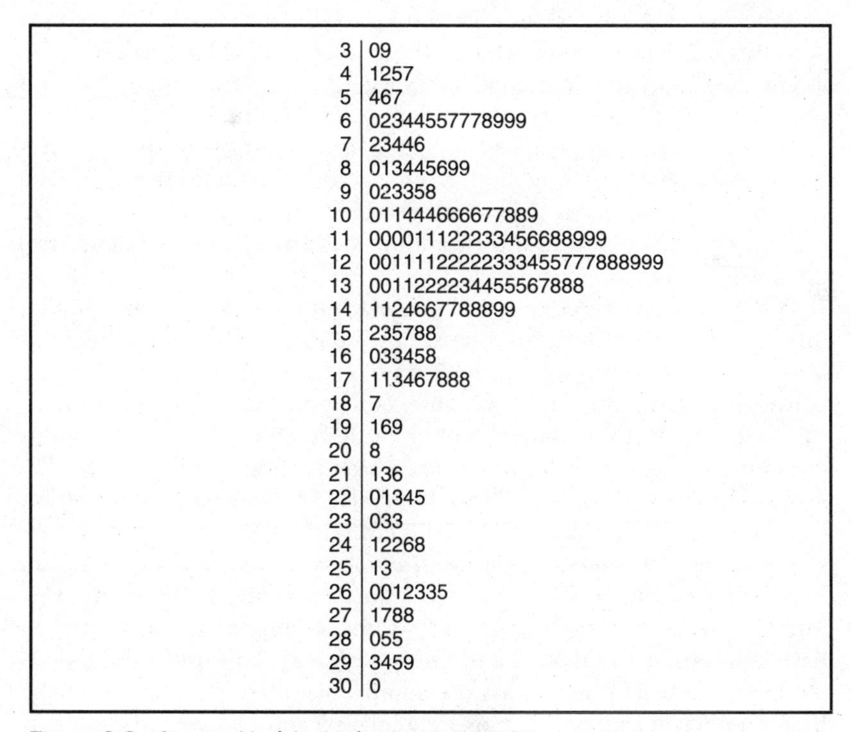

Stem	Leaf
3	09
4	1257
5	467
6	02344557778999
7	23446
8	013445699
9	023358
10	011444666677889
11	0000111222334566889999
12	001111222223334557778889999
13	001122223445556788 8
14	1124667788899
15	235788
16	033458
17	113467888
18	7
19	169
20	8
21	136
22	01345
23	033
24	12268
25	13
26	0012335
27	1788
28	055
29	3459
30	0

Figure 2-2. Stem and leaf display for data in Table 2-1.

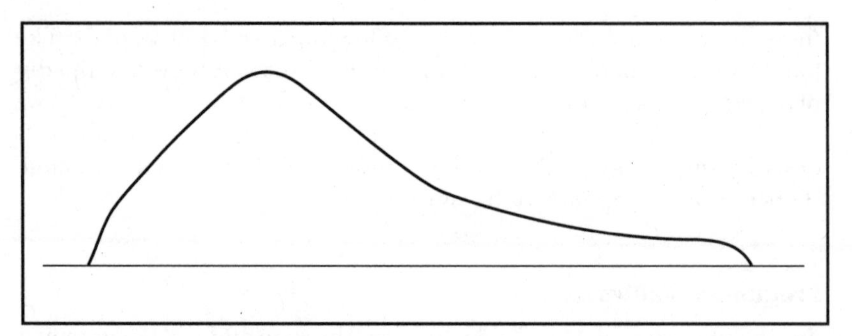

Figure 2-3. Smooth outline of the stem and leaf display of Figure 2-2.

as well have been started at a lower value.) Note that we lose the detail of the stem and leaf display. For instance, we can only tell that there were 6 values in the first interval, but their exact values are lost. From the column labeled "relative frequency" we can read that those 6 values comprised 3 percent of the entire set of observations. The relative frequency column makes it easy to see that the interval from 110 to 129 seconds accounts for about one-quarter of the phone calls (23.5 percent). The cumulative frequency column makes it easy to see that about half the phone calls (52.5 percent) lasted less than 130 seconds.

The entries under "relative frequency" are found by dividing the frequencies by 200, the total number of calls, then multiplying by 100 to express the result as a percentage. The entries under "cumulative relative frequency" are found by adding successive rows of relative frequencies.

The intervals of a frequency table are arbitrary. You can start the first interval at any value and you can use any interval width. But the intervals must be unambiguous so that there is no question about the boundaries between adjacent intervals. Each observation must fit into only one interval. In this study, for example, a phone call which is listed as 49 seconds goes into the first interval, while one which is 50 seconds goes into the second. It is not crucial that all the intervals have the same width, but for comparisons between intervals, that is convenient.

The choice of intervals is subjective. Studying the data gives you an impression of "shape," and you can adjust the initial value for the first interval and the succeeding interval widths to emphasize the shape you have in mind. In Table 2-3, for instance, we have used 7 intervals of width 50 starting at 0. This makes a more compact table which emphasizes the high number of calls which took between 100 and 150 seconds.

Table 2-2. Frequency Table for Lengths (in Seconds) of 200 Telephone Calls; 14 Intervals of Width 20

Interval	Frequency	Relative frequency (%)	Cumulative relative frequency (%)
30–49	6	3	3
50–69	17	8.5	11.5
70–89	14	7	18.5
90–109	21	10.5	29
110–129	47	23.5	52.5
130–149	32	16	68.5
150–169	12	6	74.5
170–189	10	5	79.5
190–209	4	2	81.5
210–229	8	4	85.5
230–249	8	4	89.5
250–269	9	4.5	94
270–289	7	3.5	97.5
290–309	5	2.5	100

Histograms

Graphs which correspond to Tables 2-2 and 2-3 are shown in Figures 2-4 and 2-5. These graphs are called **histograms** and are characterized by **intervals of equal width** so that the *areas* of the rectangles at each interval correspond exactly to the frequencies at that interval. This insures that the percentage of area taken up by any one rectangle is exactly the same as the relative frequency value for that interval. This in

Table 2-3. Frequency Table for Lengths (in Seconds) of 200 Telephone Calls; 7 Intervals of Width 50

Interval	Frequency (%)	Relative frequency (%)	Cumulative relative frequency (%)
0–49	6	3	3
50–99	37	18.5	21.5
100–149	94	47	68.5
150–199	25	12.5	81
200–249	17	8.5	89.5
250–299	20	10	99.5
300–349	1	0.5	100

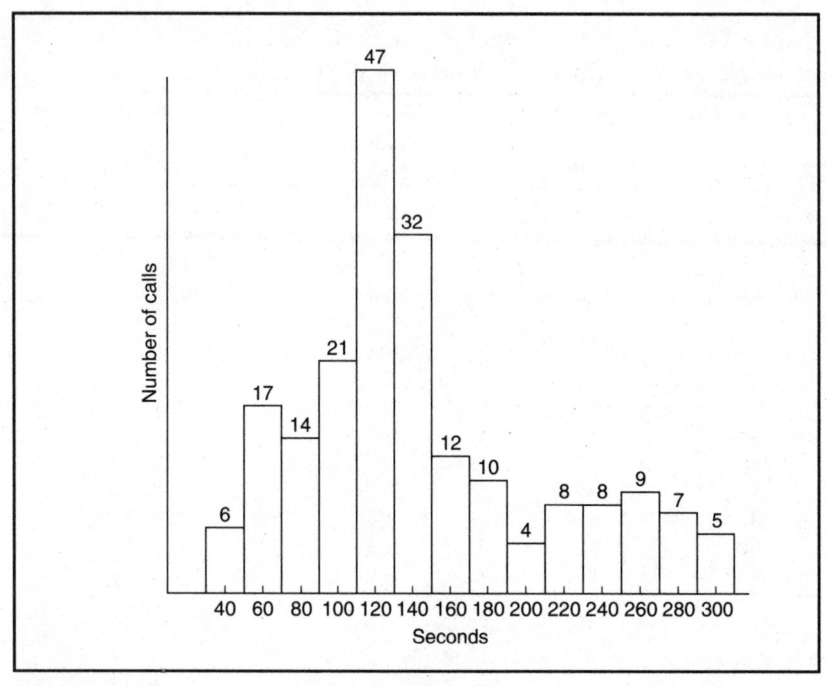

Figure 2-4. Histogram for the lengths (in seconds) of 200 phone calls. Fourteen intervals of width 20 as given in Table 2-2.

turn means that the visual effect of the graph is not distorted. If one section has twice the area as another, that is because it represents twice as many observations.

The main effect of the histogram is to give geometric shape to the data. As with the stem and leaf display, it is clear that there is a build up of calls centered around 120 seconds. The graphs are mound-shaped with a heavy tail on the right side. Such a graph or distribution is *skewed* to the right.

There is no rule for how many intervals to have in a histogram. Many authors suggest using anywhere from 5 to 20 intervals. It is up to you. Most people try several versions before they get one that looks "right," and that conveys a coherent overall impression of the pattern of the data set. Having too few intervals lumps all the data together; having too many scatters the data too thinly. Both extremes make it hard to see an overall shape. Computer programs that draw histograms have a default number of intervals which the user can change.

In drawing a histogram you must indicate the boundaries for the intervals, so that the reader clearly understands the limits for each interval.

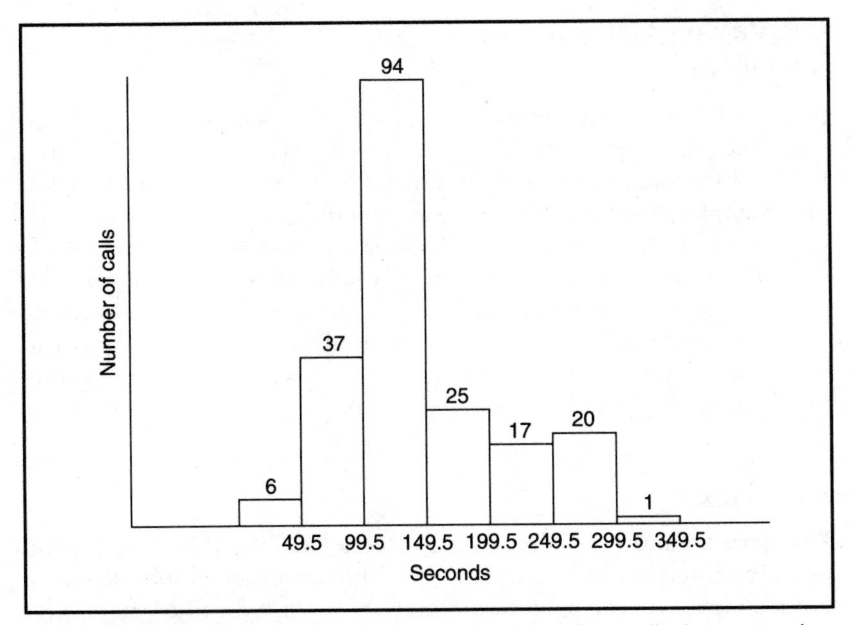

Figure 2-5. Histogram for the lengths (in seconds) of 200 phone calls. Seven intervals of width 50. The first interval starts at 0 seconds.

One standard way is to enter numbers at the boundaries of the intervals, as in Figure 2-5. These numbers are picked so that no observation falls exactly on the boundary. The boundary value is halfway between the highest number observable in one interval and the lowest observable in the next. Using the data in Table 2-3 as an example, the boundary value between the first two intervals is 49.5 seconds, which is halfway between 49 seconds (the longest possible call counted in the first interval) and 50 seconds (the shortest possible call counted in the second interval).

An alternative approach, used by many computer programs, is to label the midpoints instead of the boundaries. In this case you need to specify outside the graph how the boundaries were actually determined, so that the reader is not confused by borderline cases. Figure 2-4 is this kind of graph.

The histogram is the most popular way to display the frequency distribution of observations in a data set when the measurements are on a numerical scale. It is an essential feature that the horizontal axis is a scale which has an inherent order; the values on the left end of the axis are smaller than the values on the right. In the example here, the shortest phone calls are represented at the left end of the horizontal axis and the longest calls at the right end.

Displaying Categorical Variables

Stem and leaf displays, frequency tables, and histograms are sensible when the observations are *along an ordered scale* which runs from the lowest observation to the highest. But many studies have no such scales. For example, a study may involve counting the number of people in categories which have no numerical values, such as sex or ethnic background or brand preference. And even studies which begin with observations on a continuous scale may at some point split the observations into categories. The main purpose of graphs at this stage is to reveal the relative contributions of the various categories. This is commonly shown by using bar graphs or pie charts.

Pie Charts

Pie charts are most effective when a variable, which is easily understood as representing a "whole" amount, is split into component subcategories. For example, you can split *all* the households in the United States into subgroups by income level, or you can split your *entire* sales market into nonoverlapping regions. For a pie chart to "work," the reader must have an intuitive grasp of what the whole pie stands for and a good feeling for the comparative sizes of "slices." As a consequence they are not really very good for making fine distinctions. The facts in a pie chart can always be presented more clearly in a table.

A pie chart must also meet two mathematical requirements. First, the percentages for all the categories must total 100 percent. And second, the measures of the central angles must be in proportion to the percentages for the categories. This insures that the areas of the slices are in the correct proportions.

Example 2-4: For marketing purposes you wish to display a breakdown of households in the United States into income groups. The basic data are taken from the United States Census Bureau reports. Table 2-4 shows the figures for March 1989.

Suppose for your purposes you decide to collapse this data into five categories which you call Poor (under $10,000), Lower middle ($10,000 to $29,999), Middle ($30,000 to $49,999), Upper middle ($50,000 to $99,999), and Rich ($100,000 or more). This is shown in Table 2-5.

Even though this data was originally collected on a finer scale, at this point there are just five categories. They are not of equal width and are therefore not suitable for display as a histogram. Instead you can use a pie chart or a bar graph.

Table 2-4. United States Household Income (March 1989)

Income group	Number of households (in thousands)
Less than $5000	5,737
$5000 to $9999	10,006
$10,000 to $14,999	9,516
$15,000 to $19,999	9,126
$20,000 to $24,999	8,184
$25,000 to $29,999	7,891
$30,000 to $34,999	6,984
$35,000 to $39,999	6,414
$40,000 to $44,999	5,373
$45,000 to $49,999	4,265
$50,000 to $74,999	12,455
$75,000 to $99,999	3,966
$100,000 or more	2,911

The sizes of the angles are found by taking percentages of 360 degrees. For instance, if a category contained 20 percent of the observations, its central angle would be 20 percent of 360 degrees, or $.20 \times 360 = 72$ degrees. These computations are done automatically by computer graphics packages. The angles for the income data are shown in Table 2-6, and the chart is shown in Figure 2-6.

For a readable graph there should not be so many slices that the viewer gets lost. If you have too many small categories, the graph will be cramped. In such cases it is helpful to combine several small categories into one larger one. Remember that the main point of a pie chart is to make it easy to see the relative sizes of the categories. Try to not do anything to the chart which obscures that purpose.

Table 2-5. United States Household Income (March 1989)

Income category	Number of households (in thousands)	Percent
Poor	15,743	17.0
Lower middle	34,717	37.4
Middle	23,036	24.8
Upper middle	16,421	17.7
Rich	2,911	3.1
Total	92,828	100.0

Table 2-6. United States Household Income (March 1989)
Central Angles for Pie Chart

Income category	Percent of households	Angle (degrees)
Poor	17.0	61.2
Lower middle	37.4	134.6
Middle	24.8	89.3
Upper middle	17.7	63.7
Rich	3.1	11.2
Total	100.0	360.0

In applications where the categories have some inherent order (like incomes) the graph will be easier to interpret if adjacent slices correspond to adjacent groups. Otherwise you can arrange the slices in any order to make your point.

Pie charts are often used for comparing two or more variables, as shown in Figure 2-7, but (except for the fact that people are used to looking at pie charts) the same story could be told better using a table or a bar chart. The more "pies" there are to compare, the worse this approach is.

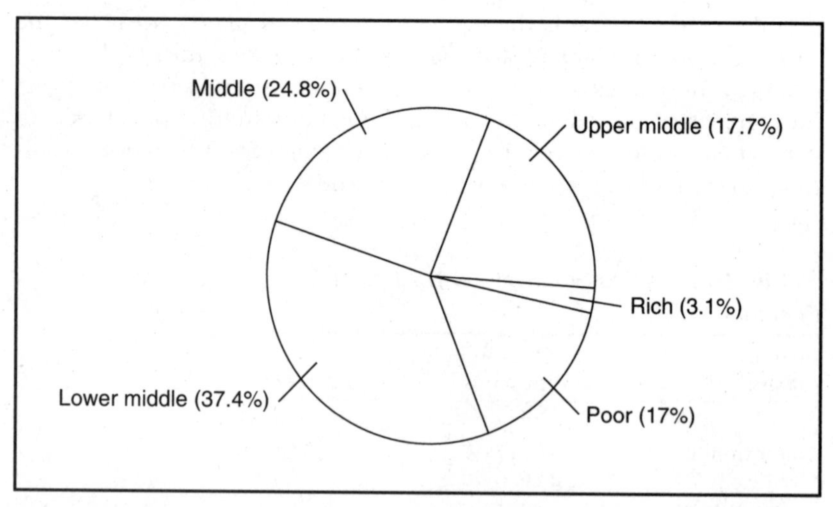

Figure 2-6. Pie chart for United States household income groups, March 1989.

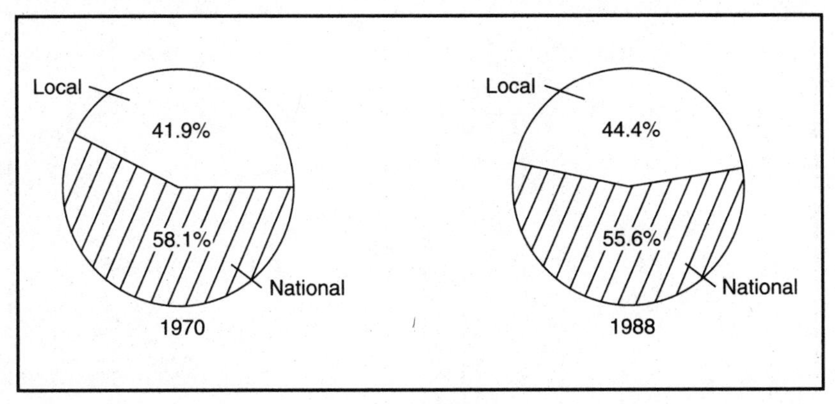

Figure 2-7. Percentage of total advertising expenditures, United States. (Source: *U.S. Statistical Abstract, 1990.*)

Bar Graphs

The **bar graph** is an excellent choice for comparing the values assigned to different categories, particularly when there is no need for any idea of a "whole" variable. The *height* of the bar corresponds to the value for the category. Otherwise the guidelines are the same as for pie charts.

In Figure 2-8 a bar graph for the household incomes is shown. Notice that both the *number* and the *percentage* of households is shown for each category. The graph is not a histogram because it does not use the fact that the income groups were originally on a continuous scale. This is emphasized by putting space between the bars. As far as this graph is concerned there are just five groups of people, and the graph just shows their relative sizes. The graph would be no different if the categories were not based on a numerical scale; this could just as well be a picture of the relative sizes of five ethnic groups.

Bar graphs are useful for comparing two or more populations on a set of variables. This is illustrated in Figure 2-9 which compares the allocation of advertising expenditures to various media at different times.

Data Maps

Many graphs make use of representational images to enhance understanding. The variety is limited only by the imagination of the artist. For example, if your company manufactures furniture you may be able to incorporate pictures of furniture into your graphs. Such graphs must

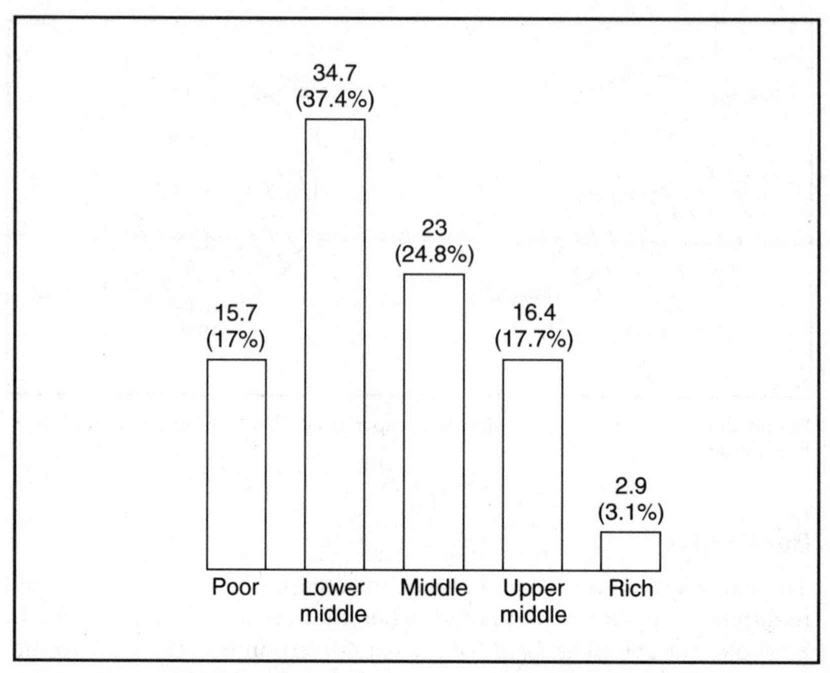

Figure 2-8. Bar graph for United States household income groups, March 1989. Numbers are thousands of households.

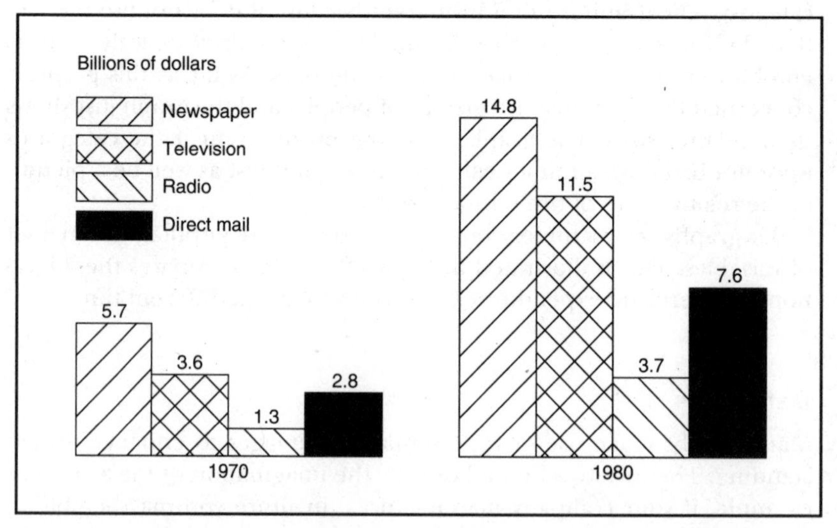

Figure 2-9. Advertising expenditures by selected media: 1970 and 1980. (Source: *Statistical Abstract of the US, 1990.*)

include the numbers represented by the pictures because the pictures themselves almost always distort the quantities they represent.

One particular style of graph, the data map, finds many uses in business when geography is a factor. These can be used to show national sales figures or targets. Different intensities of gray or color can be used to represent ranges of values. As a simple illustration, Figure 2-10 shows patterns of changing sales.

Time Series Graphs

When the horizontal axis of a graph measures time, as when you show the sales figures over a period of years, the graph displays a **time series**. A time series is a set of observations associated with equally spaced points in time (every year, every month, every hour, etc.). The usual practice is to connect the points by straight lines, ignoring any fluctuations that may

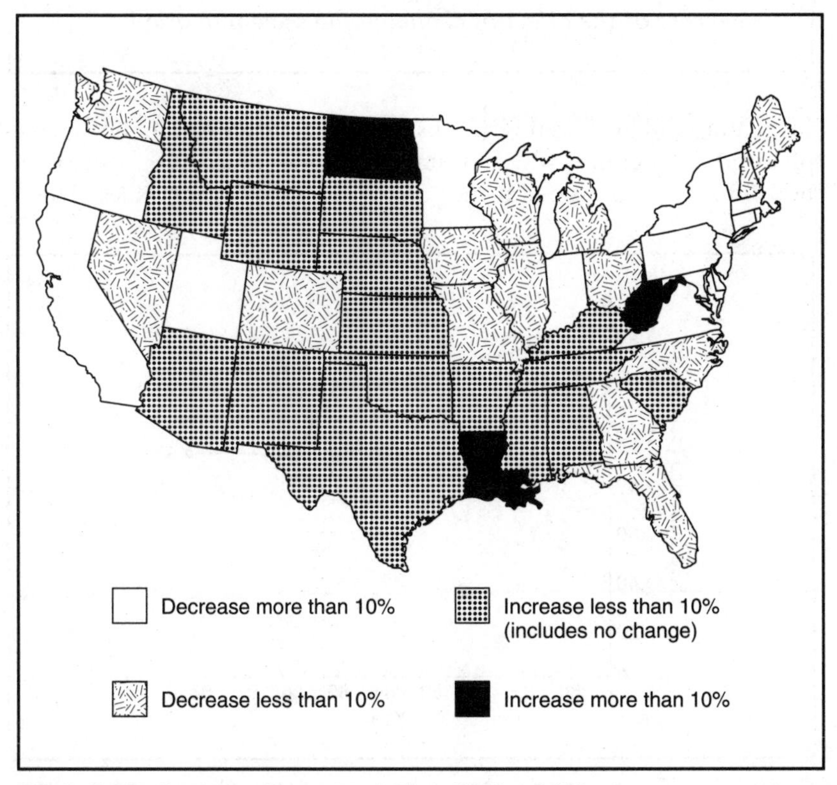

Figure 2-10. Data map. Change in sales from 1990 to 1991.

have occurred between the "official" times. In such graphs the main impression is of *change* over time. We look to see if there are upward or downward trends, sudden spikes, or cyclical repetitions.

A typical time series graph is shown in Figure 2-11. It is based on the data in Table 2-7, which were taken from the 1990 *Statistical Abstract of the US*. Such data might be used by a transportation manager interested in tracking the price of gasoline over a period of years to see, perhaps, how it influenced overall expenses or profits.

Problems with Graphs

Scale

Key Concept 7

Be aware of the effect of changing the scale of a graph.

In creating and interpreting graphs the effect of scale cannot be overemphasized. Two graphs which contain the same information can create two different impressions if they are drawn to different scales. This is

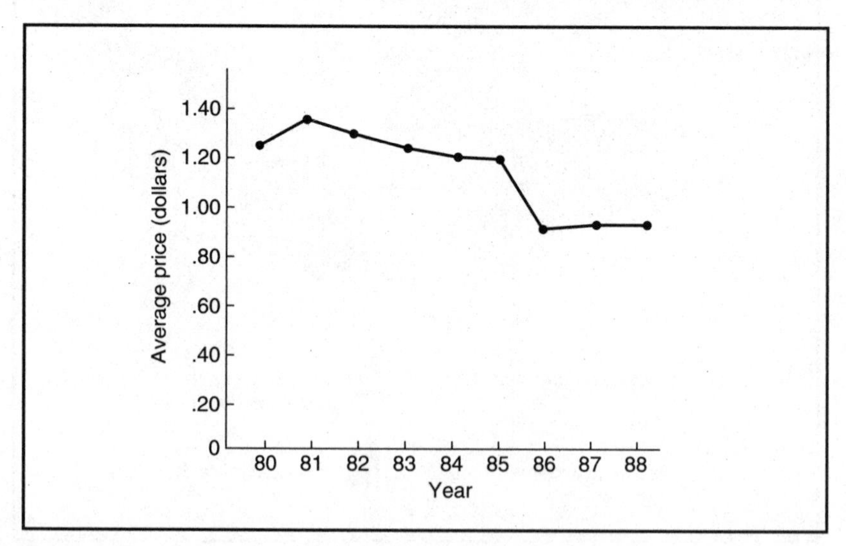

Figure 2-11. Time Series graph. Average price per gallon of unleaded gasoline in the United States.

Table 2-7. Average Price Per Gallon of Unleaded Gasoline in the United States, 1980–88

Year	Average price
1980	$1.25
1981	1.38
1982	1.30
1983	1.24
1984	1.21
1985	1.20
1986	.93
1987	.95
1988	.95

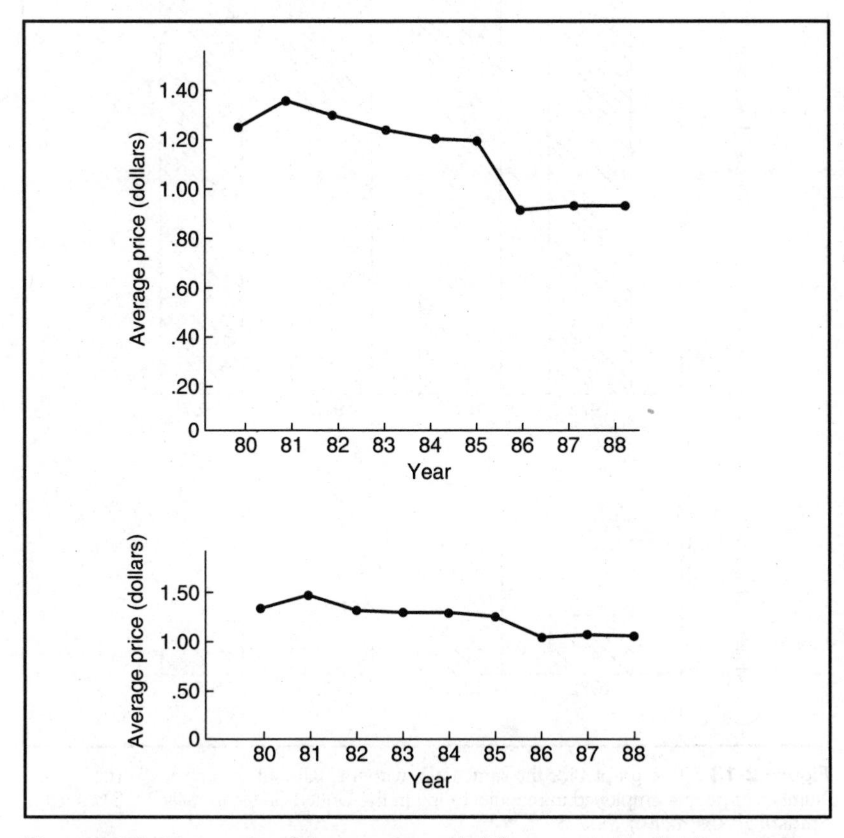

Figure 2-12. Two graphs of the same data with different vertical scales. (Average price per gallon of unleaded gasoline in the United States.)

especially true in bar graphs and time series graphs. Two categories which look about the same size on one scale may appear very different on another. To reinforce the point, Figure 2-12 shows the graph in Figure 2-11 along with another version which has a different vertical scale. You can see that steep rises and drops in the first graph appear more gradual in the second. Similar changes result from stretching or squeezing the horizontal axis.

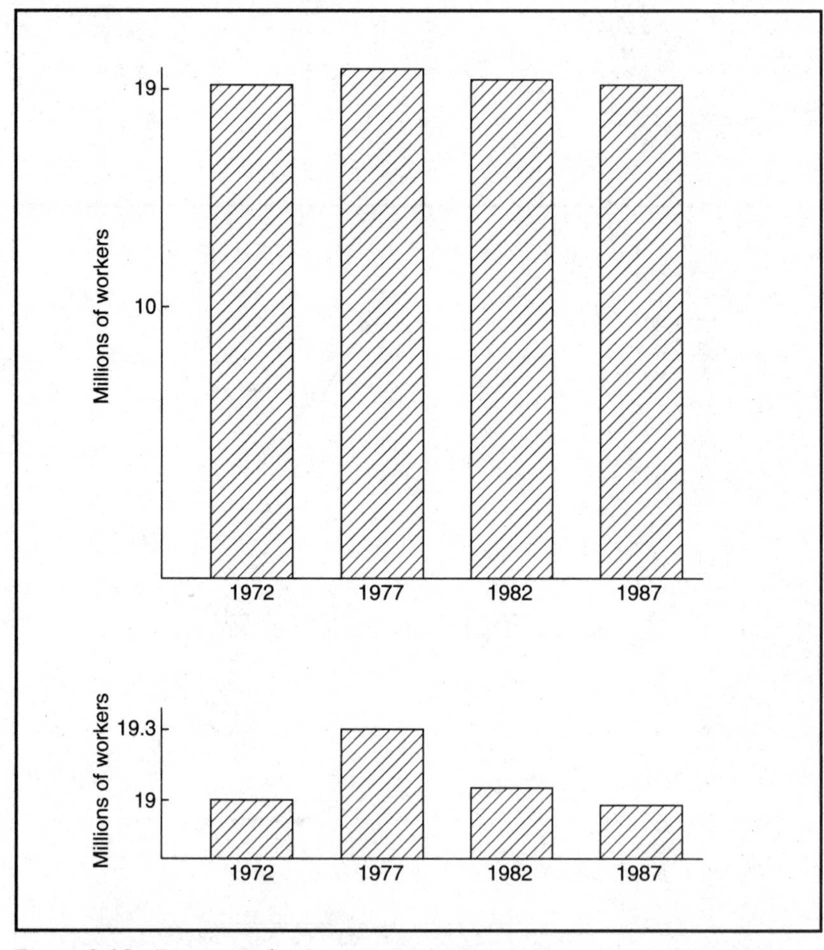

Figure 2-13. Two graphs for the same data, with and without zero on the vertical axis. Number of people employed in manufacturing in the United States in millions. (Data from *Statistical Abstract of the US, 1990*.)

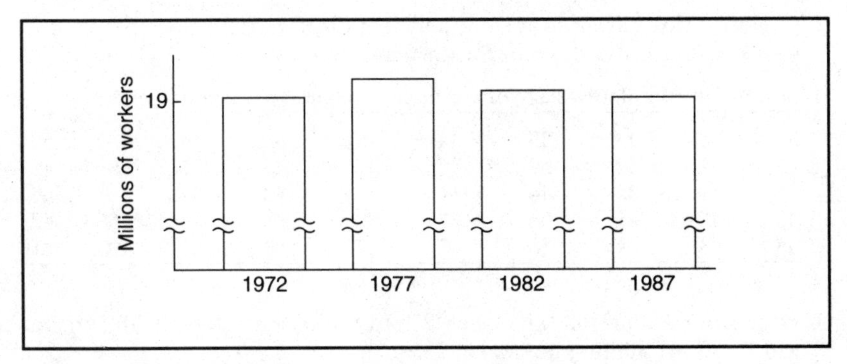

Figure 2-14. Number of people employed in manufacturing in the United States in millions. (Data from *Statistical Abstract of the US, 1990.*)

Zero

In addition to the scale itself, you must account for zero on the vertical scale. Unless the vertical axis starts at zero, your eye may be misled. For example, you expect a bar that is twice as high to represent a value that is twice as great as the shorter one. But this will only be true if the vertical axis starts at zero. Compare the two graphs in Figure 2-13 to see the effect of not starting the vertical axis at zero.

Sometimes it is unreasonable because of space limitations to show the entire vertical scale. Then it is usual to indicate a broken axis in some manner like that in Figure 2-14 which gives readers fair warning not to trust their eyes.

Self-Check

The answers to the following exercises can be found in Appendix A.

1. A department store sales manager initiated a phone survey of customers who owned the store's credit card. Included in the survey were these three questions:

   Have you made any purchases in the store in the last 3 months? If so, what was the amount of the purchase, and did you use the store credit card for the purchase?

   The responses to these three questions were recorded under the labels "Recent purchase," "Purchase amount," and "Store credit card," respectively. Classify each of the response variables as either numerical or categorical.

2. Organize this data into a stem and leaf display. Take the unit for the stems to be 10, and the unit for the leaves to be 1.

Minutes to complete a stage of assembly by a team of workers:

35	38	44	33	44	43	48	40	45	30
45	32	42	39	49	37	45	37	36	42
35	41	32	46	34	30	43	37	44	49
36	46	45	36	37	37	45	36	46	42
38	43	34	38	47	35	29	41	40	41

3. Organize the data of Self-Check 2 into a frequency table with interval width 3. Start the first interval at 27.

4. Draw a histogram which corresponds to the frequency table of question 3 above.

5. To help set the unit retail price for tangerines which were purchased by the truckload, a manager weighed some tangerines picked at random from the shipment. The histogram for the weights is shown. Read it and answer these questions.
   a. How many tangerines were weighed?
   b. What percentage of the tangerines weighed more than 5.95 ounces?
   c. What is the interval width for the histogram?

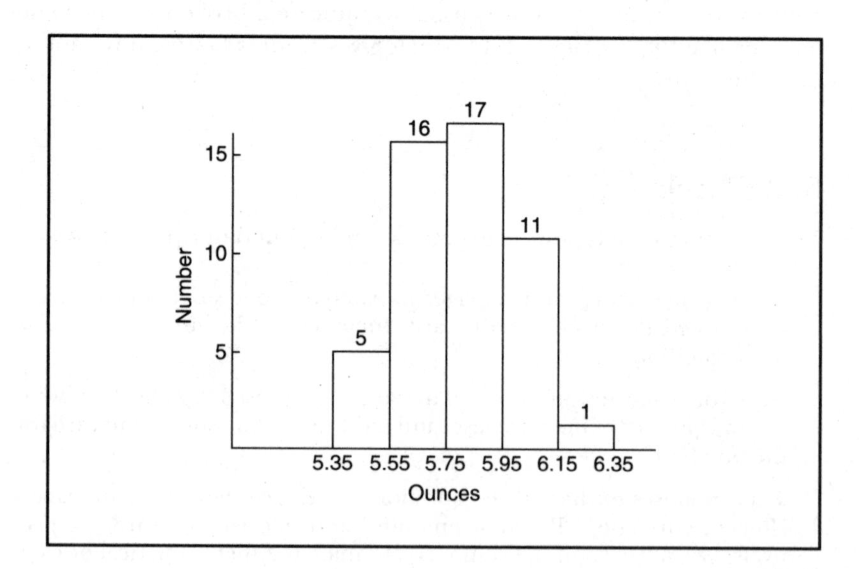

6. For a comparison of the average amount of money spent annually by men and women on 3 luxury items, which kind of graph would work best?
   a. Histogram
   b. Pie chart
   c. Bar graph

7. If the vertical scale on a time series graph were changed from 20 units per inch to 40 units per inch, would the graph look more level or more jagged?

3
Numerical Summary Statistics

Key Concepts

8. Important measures of centrality are the mean, the median, and the mode.
9. Report averages separately for important subgroups of a population.
10. A data set should be summarized by both a measure of centrality and a measure of variability.
11. An important measure of variability is the standard deviation.
12. The standardized score is a measure of relative standing in a group.
13. The normal curve, which has wide applicability in statistics, is described completely by its mean and standard deviation.
14. Ascertain the reliability and validity of the initial data before drawing conclusions from it.

The Language of Statistics

Measure of centrality

Average

Mean

Median

Mode

Skewed distribution

Quartiles

Measure of variability

Deviation from the mean

Standard deviation

Standardized score

Normal distribution

Normal curve

Validity of data

Statistical quality control

Introduction

In addition to graphical displays, a statistical report also contains useful numerical summaries. This chapter discusses the most frequently used of these, especially averages and measures of variability.

Averages—Measures of Centrality

The general public takes the word "average" to mean typical or representative. When you hear a manager say that "on average" receipts are $1000 per day, you would assume that on most days this business takes in about $1000. And this is exactly the point of using an average. The **average** condenses a lot of data (the original collection of measurements) into just one number, which is supposed to typify the middle of the data. In the technical jargon of statistics, averages are called "**measures of centrality**," "measures of central tendency," or "measures of location" for this reason. There are many ways to condense a set of measurements to get an average, but depending on the nature of the original data, the result may or may not be typical of the middle. It is therefore useful to be familiar with the properties of a few of the more commonly used measures.

Key Concept 8

Important measures of centrality are the mean, the median, and the mode.

Figuring the Arithmetic Mean

This is usually just called *the* mean, and is the most familiar kind of average. You compute the **mean** of a set of values by adding all the individual values and then dividing this sum by the number of individual values. It is a good choice as a typical middle value when the original list of values is more or less evenly distributed from the lowest values to the highest. Here are two contrasting examples.

Example 3-1: We show the list of salaries for the 10 employees of one department of a business. Find the mean salary.

Salaries department A (thousands of dollars):

18,19,20,20,20,20,20,22,24,28

Because the sum of these 10 values is 211, the mean is 211/10, which is 21.1 thousand dollars, or $21,100. This is reasonable as a "typical" middle salary.

Example 3-2: Find the mean salary in department B using the given data.

Salaries department B (thousands of dollars):

18,19,20,20,20,20,20,22,24,120

In this case the mean is 303/10 which is 30.3 thousand dollars, or $30,300. Yet, it seems strange to say that the average salary in this department is $30,300. No one in the department earns a salary close to that. For this set of data the mean is not reasonable as a representative of "middle" salaries.

What happened? Why are the results so different? The calculations are correct, but only in one case is the mean typical of the middle of the list. In the second example the mean is actually higher than all the salaries except the top one. In fact, it is exactly the top salary that caused the mean to come out as large as it did. A collection of values which includes one or more exceptionally high values is called skewed towards the high end. Conversely, a distribution which contains one or more exceptionally low values is skewed to the low end. The main point of the illustration is that for **skewed** sets of data, the mean is not representative of the typical middle value, but is pulled towards the extreme values. For such data sets the median may be a more appropriate average.

Figuring the Median

The median is an average which is not sensitive to skewness. **Median** simply means "middle"; it is numerically the middle value. In order to calculate the median, therefore, the individual values must first be put into numerical order, at which point the middle one is the median. If there are an even number of values in the list, there will be *two* middle numbers. In this case the median is taken to be the mean of those values; it is halfway between the two middle values. In general, if there are n values in the data set, then once they are in order, the position for the median is given by $(n+1)/2$. In Examples 3-1 and 3-2, each data set contains ten numbers, so that the median is the value halfway between the fifth and sixth numbers in the list; $(n+1)/2 = 5.5$. In both examples the median is 20, which is a reasonable measure of centrality.

The purpose of the median is to split the data into two equal-sized groups, with half the data above the median and half below. Notice that the median is not influenced by the one extremely high salary as was the mean. Because distributions of personal income are often skewed, statistical reports which involve incomes of large groups of people will probably present medians.

Example 3-3: Table 3-1 is an excerpt from a table in the 1990 *Statistical Abstract of the US* which is published by the United States Census Bureau and is a good source of national and regional information.

The first conclusion to be drawn from the table is that about half the new houses sold in the United States in 1988 sold for less than $112,500 and half sold for more than that. In the same way, the figure for the

Table 3-1. Median Sales Price of New, Privately Owned, Single Family Houses Sold in 1988

U.S.	$112,500
Northeast	$149,000
Midwest	$101,600
South	$ 92,000
West	$126,500

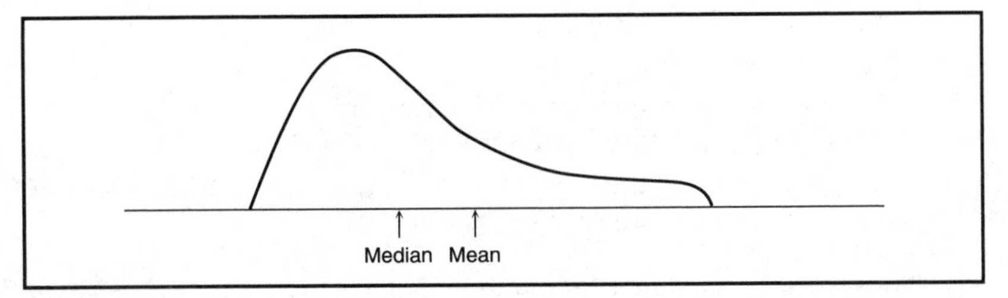

Figure 3-1. For skewed distributions the mean is pulled more to the tail than the median.

Northeast implies that roughly half the new houses sold for over $149,000 and half for less. Note in particular that the figures are quite different from region to region, and that if you were only presented the overall U.S. figure, you would be losing a good part of the picture.

The median price for house sales is reported, rather than the mean, because the distribution of sales prices is skewed by the sale of comparatively few but extremely high priced houses which would inflate the mean, as in Example 3-2. In a symmetric distribution (one which is not skewed) the mean and the median are close in value; but in a skewed distribution the mean is nearer to the end with the extreme values. This can be shown by a sketch like Figure 3-1.

Averages for Subgroups of Populations

Key Concept 9

Report averages separately for important subgroups of a population.

When you interpret any average, as is evident in Example 3-3, you should always think about the existence of important subgroups for whom the average may be very different. An average, remember, condenses all the original data into one number, and it is prudent to check that you have not lost important detail. It is quite possible to drown in water which has an average depth of two feet.

Example 3-4: These figures were reported in the *New York Times* for February 18, 1990. They were compiled from data released by the

American Medical Association. The report did not define what was meant by "average."

> The average net income (after expenses and before taxes) for all physicians in the United States in 1988 was $144,700.

But be careful about lumping all physicians together. Look at the figures broken down by the more popular specialties.

1988 average net income (thousands of dollars)	
Surgery	207.5
Anesthesiology	194.5
Obstetrics/Gynecology	180.7
Pathology	131.0
Internal Medicine	130.9
General/Family Practice	94.6

You see that the overall average lumps together some quite different groups. In any given statistical report it is up to the issuer of the report to decide when it is and when it is not reasonable to combine subgroups.

Weighted Means

It is often useful to be able to combine the means of subgroups to find the mean of larger groups. If the subgroups are not all the same size, then this computation is called a weighted mean. As an illustration, suppose a dealership sells two similar products: On product A their mean profit is $2000 per sale, while on product B the mean profit is $3000. Note that the overall combined mean profit per sale is not necessarily $2500, which would be the correct value if they sold the *same number* of each product. If they mainly sold product A, then their mean profit per sale would be nearer $2000, while if they mainly sold B, their mean profit per sale would be nearer $3000.

The computation of the weighted mean is done by "weighing" each of the contributing values by an appropriate percentage in such a way that all the weights add up to 1. The computation is expressed as:

$$\text{weighted mean} = \text{weight } 1 \times \text{value } 1 + \text{weight } 2 \times \text{value } 2 + \text{etc.}$$

If it happens that all the weights are equal, then the weighted mean is just the ordinary mean of the contributing values.

Example 3-5: A dealership sells products A and B. For the previous month 70 percent of their sales was product A and 30 percent of their sales was B. If the mean profit per sale is $2000 for A and $3000 for B, what was their mean profit per sale for the combined sales of A and B?

Solution:

$$\begin{aligned} \text{weighted mean} &= .70(2000) + .30(3000) \\ &= 1400 + 900 \\ &= 2300 \end{aligned}$$

So the mean profit per sale on all sales is $2300.

Figuring Quartiles

Related to the median, which splits the distribution into two equal-sized chunks, are **quartiles**, which split it into four equal-sized chunks. The *first quartile* is the number that splits off the lowest 25 percent of the values from the rest. The *second quartile,* which is the same as the median, splits off the lower 50 percent from the upper 50 percent. The *third quartile* splits off the highest 25 percent from the rest.

If it were necessary to split the distribution into five equal parts the division points would be called "quintiles." In the same way ten parts would be "deciles," and one hundred parts would be "percentiles."

Example 3-6: For a market analysis a research team describes the income distribution in a potential sales region. The information is presented two ways here, both of which are commonly used. They are not exactly equivalent, because one uses fixed income groups and the other uses fixed percentages, but they convey roughly the same information.

From Table 3-2 it is easy to see that 18.7 percent of the families have income in the $15,000 to $24,999 range; from Table 3-3 it is easy to see that 25 percent of the families have income less than $18,500.

Table 3-2. Money Income of Families
Percent Distribution of Families by Income Level

Income group	Percentage of population
$0–$14,999	20.7
$15,000–$24,999	18.7
$25,000–$49,999	37.7
$50,000 and over	22.9

Table 3-3. Money Income of Families
Distribution of Income by Quartiles

First Quartile	$18,500
Second Quartile	$30,800
Third Quartile	$48,200

Figuring the Mode

In some data sets it is natural to report any value which occurs more frequently than the others. This value is called the **mode** for the data set. When the data are collapsed into intervals like those of a histogram, then the interval with the highest frequency count is called the modal interval. The word "mode" simply means "popular" (or stylish, as in "pie a-la-mode"). The mode is uniquely applicable when the numerical values in a data set are really just labels for categories. For example, if you record sales of blouses by size, then the most popular size would be called the mode. It would make little or no sense to compute the mean size for an order of blouses.

Example 3-7: A food market manager might allocate shelf space based on expected sales. In this sense, the brand of soup with highest past sales would get maximum shelf space. The manager would record frequency of sales by brand. The most popular brand would be the "modal" brand.

Variability

When you are familiar with a business, you have a sense of what the relevant averages mean. If you read "average daily sales is $22,000," you have a sense of how the business is doing. But in a less familiar field such a statement can be too vague. Here's an example.

Example 3-8: Over the last four years the average (mean) profits of our company have been $1.2 million. Here are three possible scenarios that could have led to this statement.

Year	A Profit ($ millions)	B Profit ($ millions)	C Profit ($ millions)
1	1.1	0.2	2.6
2	1.3	3.0	1.4
3	1.2	0.3	0.6
4	1.2	1.3	0.2

In each case the mean of the four numbers is $1.2 million. But the three columns tell different stories. In case A, the numbers are fairly stable and close to the mean. In B, there is erratic rising and falling, and in C there is a consistent downward trend.

Trends over time are best seen in time series graphs as we have illustrated in Chapter 2. But, in addition, we can use certain standard numerical measures to describe the variability in each set of values. It is very important to realize that if you know *only* the average but nothing about variability or trend then you may seriously misinterpret the story behind the average. The next section presents a popular **measure of variability**.

Key Concept 10

A data set should be summarized by both a measure of centrality and a measure of variability.

Key Concept 11

An important measure of variability is the standard deviation.

Standard Deviation

"Deviation" in statistics means **deviation from the mean**. If the mean wage is $7 per hour, and Harold gets $9 per hour his deviation is +$2. If Anders gets $5.50, then his deviation is –$1.50. The plus and minus signs indicate above and below average. Mathematicians have worked out a formula (given in Appendix B) for computing a representative value for the deviations in a data set. The result is called the **standard deviation**, and it is derived from the average of the squares of the deviations of all the values in a data set.

When the observations in a data set are quite similar to one another they will have relatively small deviations from the mean, and so the standard deviation for the data set will be relatively small. The extreme case would be a data set in which the observations were identical; then all the deviations would be zero, and the standard deviation would be zero. In contrast, if the observations are widely scattered, then the value for the standard deviation will be relatively large. In summary, the standard deviation is a number that represents the amount of variability in a data set. Larger values represent more variability.

Example 3-9: You are the purchaser of timing devices which are used to detonate explosives in the blasting of new roadways. You must choose between two suppliers, call them A and B. In the specifications you see that the fuses sold by A have a mean time to detonation of 30 seconds with a standard deviation of 0.5 seconds, while those from B have a mean of 30 seconds with a standard deviation of 6 seconds. It is clear that those from A are less variable in their behavior, and are therefore safer to use.

Using the Standard Deviation to Give the Relative Placement of Individual Values in a Data Set—Standardized Scores

Key Concept 12

The standardized score is a measure of relative standing in a group.

Statisticians often give the placement of individual values in a data set in terms of standard deviations. For example, in addition to saying that production by John Jones is so many *units* below average (based on all production workers), the statistician might also say that his production is so many *standard deviations* below normal. This number is called the **standardized score** for Jones's production. It is calculated by comparing his deviation to the standard deviation.

Example 3-10: Suppose as part of a management project we measure the time it takes to serve customers at a bank once they reach a teller. We record the time for several hundred customers and then compute the mean and standard deviation. Suppose it turns out that the mean transaction time is 3.2 minutes, with standard deviation equal to 1.3 minutes. We can use this information to describe the placement of a transaction that took 5.8 minutes. We can compute the standardized score for this transaction.

Because the time for this transaction was 2.6 minutes longer than the mean transaction time, its deviation is 2.6 minutes. This is twice the standard deviation. The calculations are shown here.

$$\text{deviation} = \text{observation} - \text{mean} = 5.8 - 3.2 = 2.6 \text{ minutes}$$

$$\text{standard deviation} = 1.3 \text{ minutes}$$

$$\text{standardized score} = \frac{\text{deviation}}{\text{standard deviation}} = \frac{2.6}{1.3} = 2$$

From Example 3-10 you can see that if the deviation for an observation is *twice* the standard deviation, then its standardized score is 2. If the deviation is *three times* the standard deviation, then its standardized score is 3, and so on. A positive sign is attached for observations which are above average, and a negative sign is attached for observations which are below average. Thus, if you read that an observation had a standardized score of −1.5, you know that it was one and a half standard deviations below the mean.

Example 3-11: Once again, consider the study in Example 3-10. The mean time to serve a customer was 3.2 minutes and the standard deviation was 1.3 minutes. What is the standardized score for a transaction which took 2 minutes?

$$\text{standardized score} = \frac{2 - 3.2}{1.3} = \frac{-1.2}{1.3} = -0.92$$

This standardized score tells you that the observation was below the mean and that its deviation was about nine-tenths of a standard deviation. In other words, the transaction took a little less time than the average transaction. Note that if an observation has exactly the same value as the mean, then because its deviation is zero, its standardized score is *zero*. In many statistical reports the standardized score is represented by the letter "z" and is called the "z-score."

In practice, as a rule of thumb, any observation two or more standard deviations from the mean merits further investigation, because in most applications it is rare to see an observation that far from the mean. The reason for this will be explained when we discuss the so-called "normal" distribution, but meanwhile it is a useful rule to remember.

The Standard Deviation and Quality Control

Because the standard deviation measures the consistency of the observations in a sample, it has a role to play in quality control. Statistical quality control (often also called statistical process control) is the body of statistical techniques used to monitor the quality of the output of a production process. Much of the success of Japanese manufacturing after World War II is credited to their very serious adoption of these techniques under the advice of W. E. Deming, perhaps the most influential American statistician of our time in the area of business management. The next example illustrates the use of the standard deviation in setting control limits to determine when a production process may be going out of control.

Example 3-12: An established production line produces ball bearings with a mean diameter of 75 millimeters and a standard deviation of 0.1 millimeters. Suppose that every 15 minutes a random sample of 25 bearings is inspected, and the mean diameter recorded. Using statistical theory and *incorporating the 0.1 millimeter standard deviation as a measure of the variability inherent in the production process,* it can be determined that it would be very rare for this line, when it works properly, to yield a random inspection sample whose mean was less than 74.96 millimeters or more than 75.04 millimeters. Thus, if it happens that a sample mean falls outside these control limits, that would be cause for immediate investigation. It is likely that something has gone wrong. (The formula for this calculation is given in Appendix B.)

The Normal Distribution

Key Concept 13

The normal curve, which has wide applicability in statistics, is described completely by its mean and standard deviation.

One particular graph occurs so often in statistical studies that it is worth commenting on now, because interpreting it properly involves using both the mean and standard deviation. This is the so-called "normal" or "bell" curve; in engineering applications it may be called a "Gaussian" curve. Its familiar shape is shown in Figure 3-2.

Mathematically speaking, the exact smooth shape of the normal curve is determined by a fairly complex equation (given in Appendix B), and as a consequence, no *actual* set of data can be *exactly* normally distributed. However, many data sets are close enough for practical purposes and when we say that they are "normal," we really mean that they are *approximately* normally distributed.

Normal distributions of data typically result when a variable is the con-

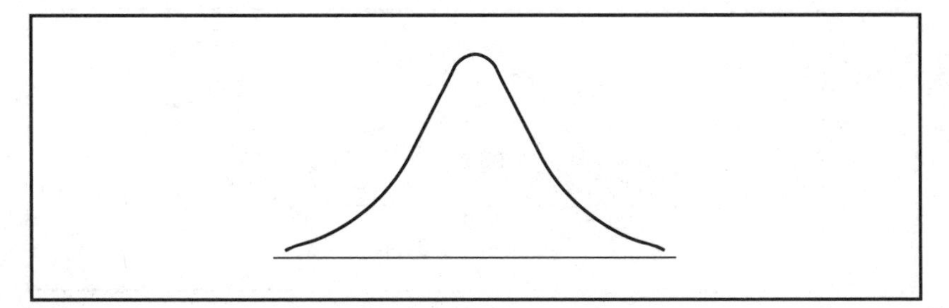

Figure 3-2. The normal curve.

sequence of many contributions which are subject to small random disturbances. Some examples of variables which are often found to be approximately normal are human traits and abilities (such as height, weight, and aptitude test scores in some population), and physical properties of manufactured goods (such as the diameter of ball bearings, or the gas mileage for a certain model automobile).

Normal Curves and the Standard Deviation

When the shape of a distribution is close enough to be called approximately normal, certain properties of the theoretical normal curve can be used. For example, for any set of data that has a normal distribution it will follow that about 68 percent of the observations are within one standard deviation of the mean, and that about 95 percent are within two standard deviations of the mean. (That is why it is rare to find an observation for which the standardized score is more than 2.) These specific percentages characterize a normal distribution and are shown in Figure 3-3. More detailed descriptions of the proportions of observations in various intervals are recorded in normal curve tables. One such table is given in Appendix C.

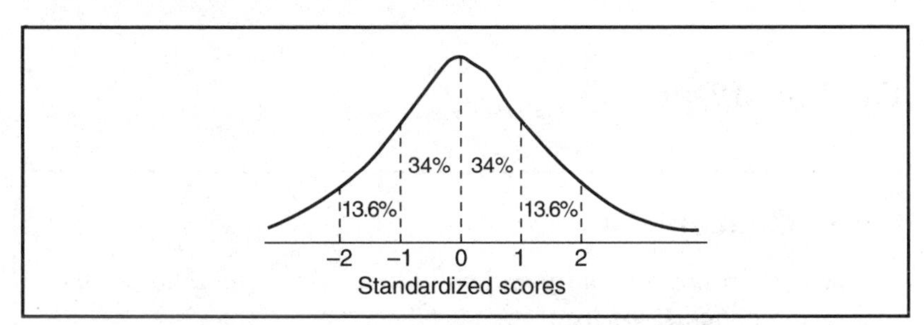

Figure 3-3. Proportion of area in some intervals of the normal curve.

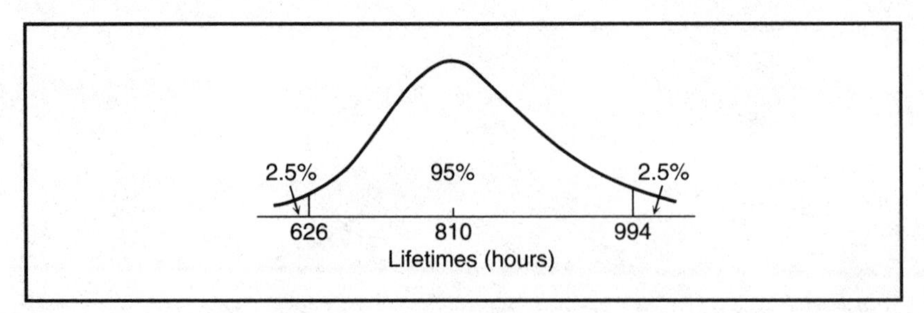

Figure 3-4. Normal approximation of failure times for light bulbs with mean 810 hours and standard deviation 92 hours.

Example 3-13: A manufacturer of light bulbs tests a new product by letting several hundred of the new bulbs burn until they fail. Even though they were all built to the same specifications they do not fail at the exact same time. There are unaccountable small variations in material and assembly. As a consequence, the histogram for the failure times will be approximately normal. Thus we can say that the lifetimes are (approximately) normally distributed, and we can accurately describe the pattern by reporting just the mean and the standard deviation.

Suppose that calculations based on the lifetimes of the tested bulbs show that the mean failure time for these several hundred bulbs is 810 hours and the standard deviation is 92 hours. If we can assume that this sample is representative of the entire production of these bulbs, then we can infer that about 68 percent of the entire production of bulbs will fail between 718 hours and 902 hours, and that about 95 percent of the bulbs will fail between 626 and 994 hours (see Figure 3-4).

This information will be important in the sales and advertising of the bulb. For instance, if the manufacturer guaranteed the bulbs for 626 hours, it would be predicted in advance that about 2.5 percent would fail before the guarantee. If this is not acceptable, then the normal curve can be used as a guide for setting another guarantee period. In similar fashion, as we show in Chapter 5, the normal curve plays a crucial role in making other estimates and predictions based on sample data.

Quality of Data

Key Concept 14

Ascertain the reliability and validity of the initial data before drawing conclusions from it.

Are the Data Recorded Correctly?

Until now, it has been assumed that the data used to construct the tables and graphs in Chapter 2 and the measures of this chapter are correct. By "correct" we mean that the numbers were accurately recorded, and that, in fact, they were the appropriate observations to make in the first place. If you are the person responsible for the collection of data, then you must take pains to see that both issues are addressed. Nothing is more embarrassing than to make a presentation with marvelous graphics and then be unable to answer a question such as, "Where did that peculiar piece of data come from?" or "How could you have a phone call that lasted negative 3 seconds?" Always have a colleague (someone not as immersed in the data as you) examine graphs and tables to look for nonsense or questionable data. This is particularly important with computer-generated displays. If any piece of data is incorrectly entered, a computer program may not catch it, but simply incorporate this wrong value into the output. Perhaps the most common error is the misplaced decimal point, recording a sale of $23,000 as $230.00, for example. Another error is to have blanks interpreted as zeros, so that a transaction which resulted in "no sale" gets reported as a transaction for which there was a sale of $0.00. This will grossly distort the mean sale price and other statistics based on the mean sale price.

Where Do the Data Come From?

If you are using data collected by others, especially those outside your own organization (sometimes called "external" data), find out how it was collected. Particularly find out if it is, in fact, representative of the population you are attempting to describe.

Case Study:

SAT Scores

In 1990 the United States business community expressed great unhappiness with the SAT scores of that year's college applicants, and there was much discussion about what was wrong with American education. (See *Newsweek,* September 10, 1990.) Programs were started in several states to involve business leaders in discussions about improving education.

A first task was to identify those states that had the highest SAT scores to see what they were doing right. It was clear that the midwestern states did well. What was it about education in those states that could account for the difference? This is a complex issue worthy of serious discussion, but for our immediate purpose, the first question is "Where did the data come from?" Who took those tests?

It turns out that in the midwest, only the very best prepared students (generally speaking), who want to apply to the toughest universities in the east and the west, take the SAT tests. Most colleges in the midwest use a different standardized test (the ACT) for admissions decisions, and so the majority of the college bound take the ACT instead of the SAT. In short, the interpretation of high SAT scores from the midwestern states is clouded (or confounded) by the fact that the group of students who take them is not typical. We are therefore unlikely to learn *anything* of interest about differences between the educational systems of the midwestern states and the others.

The population of interest is *all* college bound students from the midwest, but the population that was actually sampled was only the *best prepared* college bound students there. You need to be especially careful about this kind of problem when you are using external data; you may be tempted to *assume* that someone else's study meets your needs.

Does the Data Mean What You Think It Means?

A potentially more difficult problem with data is usually called the problem of "**validity**." It asks the question, "Are you measuring the right thing in the first place?" Suppose someone shows you a questionnaire to be administered to your employees, which is supposed to measure creativity. How do you know it works? You want to see evidence of its validity.

Case Study

Validity of Questionnaire

In 1988 a private study was launched to determine the characteristics of hospital staff nurses which affect the likelihood of their quitting the hospital. Several hundred nurses selected randomly in the New York metropolitan area were asked to respond to specially designed questionnaires. Within the questionnaire was a series of questions which taken together were supposed to indicate how a nurse responded to "authoritarian" management. Another series of questions was supposed to indicate whether these nurses preferred to work "independently" or with more supervision. These and other components of the questionnaire were taken from "the established literature" in the general field of human behavior. But unless these personality scales had been devised and tested on a very similar group of nurses, it is risky to assume that a nurse's score on this scale means anything at all.

There is no easy solution to this problem, especially if you are breaking new ground. One good approach is to study the results of similar projects and to get advice from others who have tried similar projects. This is the statistician's version of the famous GIGO problem, GIGO being the acronym for "Garbage in, garbage out." Probably the best advice is to get qualified help when the demands of your project push you past the point where your own experience gives you confidence. This is particularly true in the design and administration of surveys and questionnaires.

Self-Check

Answers to the following exercises can be found in Appendix A.

1. When a set of data is highly skewed, which kind of average is more representative of the center of the data?
   a. the mean
   b. the median
2. True or false? About half the values in a set of data are less than the mean.
3. Garrison Keillor claims that in the town of Lake Wobegon "all the kids are above average." Which of these comments are correct?
   a. The claim could be right if the kids were being compared to the average of kids from other towns.
   b. The claim can't be true if you are comparing them to the average of just Lake Wobegon kids, because the average is always somewhere between the highest and the lowest scores.
   c. This is clever humor. Don't spoil it by turning it into a statistics problem.
4. Why might the mean price for new home sales in a city not be a good average to use when you want to give an idea of what is typical?
   a. It is possible that a few new houses were exceptionally high priced, which would cause the mean to be higher than the typical price.
   b. It is too hard to compute.
5. If your business is supplying numerals for football jerseys, which of these measures would be of more use to you: the mean, the median, or the mode of the numerals you sell?
6. For this distribution of data, which would be higher: the mean sales price or the median sales price?

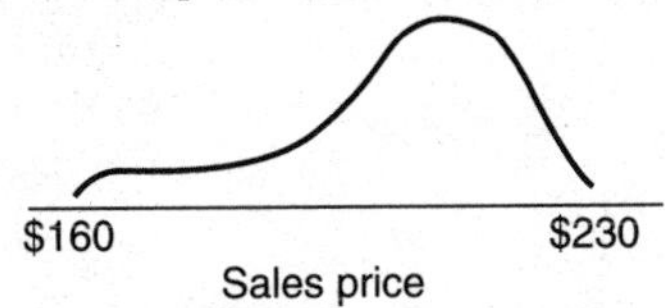

7. Which of these sets of data would have the larger standard deviation?
   a. 300, 301, 302, 303, 304, 305
   b. 10, 50, 100, 110, 200, 250

8. True or false? If the mean of a data set is 450 then all of the observations are around 450.

9. In a recent debate one politician claimed that things were getting better because the average income in the United States had gone up over the past four years. Her opponent said things were getting worse because there was a wider gulf between the average income of the rich and poor. Could both statements be right?

10. You are thinking of distributing a line of soft-drink dispensing machines. The machines are supposed to drop eight ounces of liquid into a cup. You ask two competing manufacturers to report the results of lab tests which involve using their machines 1000 times and recording the amount of liquid released each time. When you look at their specifications you see the following figures.

    Company A: Mean = 8.001 oz
    Standard deviation = .1 oz
    Company B: Mean = 8.001 oz
    Standard deviation = .6 oz

    Which company produces a more reliable dispenser?

11. The employees of a certain company are paid as shown below.
    Group 1. 1000 people whose mean wage is $6.00 per hour.
    Group 2. 200 people whose mean wage is $10.00 per hour.
    a. Show that the overall mean wage for all 1200 people is $6.67 by finding the total amount of money paid for all of them to work one hour and dividing that sum by 1200.
    b. Suppose group 1 gets cut to $5.00 per hour and group 2 gets raised to $15.00 per hour. What would the new mean wage be? Suppose it was announced that for the average employee nothing had changed. Would that be correct?

12. New employees are given two aptitude tests. On test A the mean score is 100 with a standard deviation of 15. On test B the mean score is 400 with a standard deviation of 50. Pat scored 115 on test A and 425 on test B. On which test did Pat do better compared to average? Find the answer by computing Pat's standardized score for each test.

13. A production line is considered "out of control" when its hourly output falls more than two standard deviations below the mean hourly output for the previous day. For this line the mean hourly output yesterday was 370 pieces with standard deviation 5 pieces per hour. Here are the outputs for the first several hours of the day. At what time was the line considered out of control?

Time:	8A.M.	9A.M.	10A.M.	11A.M.	12N.	1P.M.	2P.M.
Output:	369	367	365	363	361	359	357

14. The lifetimes of a new line of computer batteries is approximately normally distributed with mean equal to 42 hours and standard deviation equal to 6 hours. The batteries are guaranteed: money back if it fails before 36 hours. About what percentage of the batteries will fail before the guaranteed time?

15. The mean salary of the male managers at a corporation is $30,000 and the mean salary for female managers is $24,000. If 80 percent of the managers are female and 20 percent are male, use the weighted mean to find the mean salary for all of the managers.

16. In the graph shown in Self-Check 5 of Chapter 2, is it true that the median weight must be a value between 5.75 and 5.95 ounces?

4
Rates and Probability

Key Concepts

15. The rules of probability provide the basis for statistical inference.
16. The probability of an event is the relative frequency with which it would be expected to occur. When this is based on past observations, it is called empirical probability. When based on a hypothetical model, it is called theoretical probability.
17. Subjective probability is a measurement of one's gut feeling about the likelihood of an event.
18. The odds in favor of an event is the ratio of the probability the event will happen to the probability it will not happen.
19. The conditional probability of an event is the probability that it will occur given certain specific preconditions.
20. A tree diagram is a way of displaying sequences of events which simplifies the calculation of probabilities.
21. The Multiplication Rule is a formal calculation procedure for finding the probability of a sequence of events.
22. The Expected Value of a random variable is its mean value over the long run.
23. Vital statistics and demography are useful in long range planning.

The Language of Statistics

- Relative frequency
- Empirical probabilities
- Rate
- Theoretical probability
- Subjective probability
- Odds
- Complementary events
- Random experiments
- Equally likely
- Conditional probability
- Independent events
- Tree diagrams
- Multiplication Rule of Probability
- Expected value
- Vital statistics
- Demography
- Focus groups

Introduction

Key Concept 15

The rules of probability provide the basis for statistical inference.

The applications of statistics are vitally connected with probability. Whenever an inference is made from a partial sample to an entire population, it must be couched in the language of probability. Thus, "On the basis of first quarter sales, we will *probably* see annual profits rise." Or, "Since 4 out of 500 electronic widgets failed in the laboratory, we "expect" (meaning we assign high *probability* to) the failure rate in the field to be less than 1 percent." Because of this connection between probability and statistics, it is a good idea to have a clear understanding of some fundamental concepts of probability. That is the purpose of this chapter.

Key Concept 16

The probability of an event is the relative frequency with which it would be expected to occur. When this is based on past observations, it is called empirical probability. When based on a hypothetical model, it is called theoretical probability.

Empirical Probability

The most common interpretation of probability is that of **relative frequency**. When you hear, "There is a 50 percent probability (chance) that a new small business will go broke in less than 3 years," this means that it has been observed that 50 percent of such businesses have, *in fact,* gone broke in that time period. This percentage is taken to be an estimate of what we expect the proportion to be for the indefinite future. Such probabilities, *based on actual observations,* are called **empirical probabilities**. The crucial point in applying an empirical probability to a new event is the assumption that present circumstances are essentially the same as those on which the relative frequency was computed.

In some circumstances it is more natural to express a probability as a **rate**. Rather than say that there is a 50 percent probability that an individual new small business will fail within three years, we might say that the *failure rate* for new small businesses is 50 percent. "Probability" is more likely to be used when the emphasis is on the individual outcome, and "rate" is more likely to be used when the emphasis is on the process.

Example 4-1: A bottling company sets up its machines to fill juice bottles marked 64 ounces. In their own plant an inspection of 500 filled bottles sampled at random reveals that two bottles contain less than 64 ounces of juice. This is due to the inherent variability of the process. If the system is left unchanged, what is the probability that one of their bottles picked off the shelf at random will contain less than 64 ounces of juice?

Solution: Under the critical assumption that the system will not change, the fraction of bottles containing less than 64 ounces will continue to be about 2 out of 500. This is expressed as a percentage as follows:

$$2/500 = .004 = 0.4\%$$

The probability then is 0.4 percent, or somewhat less than one-half of one percent.

Theoretical Probability

It is also possible to assign probabilities by calculating **theoretical** relative frequencies based on a description of some ideal process. For example, given the rules of a direct mail contest used in a national advertising campaign, you could compute the probability that the winner will come from New York state. Suppose, for example, that 20 percent of the solicitations are mailed to New Yorkers. We can say then, that if response *rates* are the *same* in all states, there is a 20 percent probability that the winner will come from New York. (To say that the response *rates* are equal means the same *proportion* of people respond in *each* state.) Notice that this theoretical probability is computed before the actual event even occurs; no observations were necessary.

In making business decisions based on theoretical probabilities, you must pay close attention to any simplifying assumptions that have been made, because the further reality departs from these assumptions the less reliable is the probability figure. The mathematics of probability is quite precise, but it is rare that reality is as simple as the models on which the calculations are based. For example, it may not be true in the contest just described that response rates are the same in New York and Florida. This could happen, perhaps, if there were large differences in job status or income of ad recipients in these two states, either of which might influence motivation to participate in a direct mail campaign.

Case Study

Volkswagen Guarantee

In the summer of 1990, Volkswagen announced that it would soon offer a money back guarantee on its Passat sedan. In the United States a buyer would be able to return the car with (almost) "no questions asked" within the first three months or 3000 miles of ownership. Clearly, there is some risk involved in such an unprecedented offer. How can one estimate the probability that a buyer will return one of these cars? This is clearly a relative frequency issue, but because it is a new experiment there are no data on which to make an empirical count. One cannot count the percentage of cars returned in the past. Therefore, some theoretical basis must be found. In fact, in this case Volkswagen used a surrogate count. They used a survey of current Passat owners, which showed that 97 percent were satisfied with the car. This suggests that about 3 percent of the cars will be returned.

Whether this is a good surrogate remains to be seen. It is quite possible that the people who responded to the survey differ from the potential new buyers. For example, once people have an investment in a car, they may be more likely to rate it satisfactory. Furthermore, the owners who were surveyed were not given the option of getting their money back. This is an example of how reality is more complex than the model. The model is simple. It just says that the proportion of returns will be about the same as the proportion of current owners who express dissatisfaction with the Passat. Volkswagen was evidently willing to balance the risk of losing money on the deal with the chances of gaining a larger share of the United States market.

Subjective Probability

Key Concept 17

Subjective probability is a measurement of one's gut feeling about the likelihood of an event.

It is also legitimate to use the vocabulary of probability to describe your gut feeling about the chances of an event occurring. You may not be able to put a precise number to your feeling, but it is understood that the closer your value is to 100 percent, the more confident you are that the event will occur. You may only care to say that a probability is "low" or "high." As long as you make it clear that there is a **subjective** base to your announcement, no one will be misled. You may, of course, have a harder time defending your sense of what the correct probability is, or getting other people to invest their money based on your intuition.

Example 4-2: An investment advisor says that there is a 90 percent probability that the price of oil will go up if there is a change of government in country A. This clearly cannot be a precise probability; it must be taken only to mean that the advisor is quite convinced it will happen. You must be comfortable with the risk involved in believing the claim before taking any action based on it.

Writing Probabilities

The three interpretations of probability given above were all expressed as percentages. Careful thought will suggest that these percentages can only range from zero percent to one hundred percent. The intuitive sense of an event with a zero percent probability is that it just will not occur. Conversely, an event with a one hundred percent probability is certain to happen. For the purposes of this text you can treat "zero probability" and "impossible" as synonymous. Similarly, "one hundred percent probability" and "certain" will be synonymous. (A more advanced treatment of probability would point out that this is not strictly true, but for our purposes, it won't cause any trouble.)

When writing probabilities, you have a choice between percentage, decimal, and fractional notation. For example, you can say that the probability of event A is 25 percent, or 0.25, or 1/4 (one in four). These are equivalent, and you simply choose the one you prefer for the application at hand.

Corresponding to the statement above, that any probability must be a value from zero to one hundred percent, it follows that in decimal or fractional notation the value of a probability must always be some value from 0 to 1. The closer the probability is to 1, the more likely the event is to occur.

When switching notation from percentage to decimals, be especially careful with small probabilities. For instance, 3 percent is the same as 0.03, not 0.3. Recall that the word "percent" or the symbol "%" both mean "out of 100." Thus, 3 percent means 3 out of 100. In the same way, .03 percent means .03 *out of 100* (which is less than 1 out of 100). If you mean 3 out of 100, you write .03 or you write 3 percent or you write 3%, but *not* .03 percent.

Example 4-3: You purchase 5 out of the 10,000 raffle tickets sold in a community fund-raiser. Express the probability that you will win the one grand prize.

Solution: Your probability of holding the one grand prize ticket

$$= \frac{\text{number of tickets you hold}}{\text{total number of tickets}}$$

$$= 5/10{,}000$$

$$= 0.0005 = 0.05 \text{ percent}$$

Note: If you buy no tickets, the probability that you win is 0/10,000, which equals zero percent. Thus, "You can't win if you don't play." Conversely, if you buy all 10,000 tickets, the probability that you win is 10,000/10,000 = 1, or 100 percent, implying that you will win for sure. Many people do not realize, for example, that you can guarantee winning a typical state lottery simply by buying every possible combination. It is just prohibitively expensive, not to mention practically impossible, to fill out the forms—and, most important, you may end up winning less than it cost you to play.

Odds

Key Concept 18

The odds in favor of an event is the ratio of the probability the event will happen to the probability it will not happen.

The information given by a probability figure can also be expressed in the form of **odds**, but you have to be careful to distinguish between the odds "in favor" of an event and the odds "against" the event. You get odds by dividing one probability by another.

Example 4-4: Suppose the probability that a project will succeed is 80 percent.

a. What are the odds in its favor?

b. What are the odds against it?

Solution:

a. Since the probability that the project will succeed is 80 percent, the probability that it will fail is 20 percent. The two possibilities, "success" and "failure" are called **complementary** events. Each is the complement of the other. Two events are complementary if the occurrence of one is equivalent to the nonoccurrence of the other. The sum of the probabilities of two complementary events is always 100 percent (or 1.00 in decimal notation).

$$\text{Odds } \textit{in favor} \text{ of success} = \frac{\text{probability of success}}{\text{probability of failure}}$$

$$= .80/.20$$

$$= 4$$

This would usually be expressed by saying that the odds in favor of success are 4 to 1, since 4 is the same as 4/1. This means that for every 4 projects like this which succeed, 1 fails. Note that this is equivalent to saying that 4 out of 5 such projects (80 percent) succeed.

b. Find the odds *against* the success of the project.

$$\text{Odds } \textit{against} \text{ success} = \frac{\text{probability of failure}}{\text{probability of success}}$$

$$= .20/.80$$

$$= 1/4.$$

The odds against success are 1 to 4. This means that for every 1 failure there are 4 successes, which is equivalent to saying that 1 out of 5 such projects (20 percent) fail.

The Mathematics of Probability

The computation of the probability of events ranges from quite simple to extremely complicated. It is not too hard to see that your chances of winning the grand prize in a raffle in which you have purchased 5 out of 10,000 tickets sold is 5/10,000. But computing the probability that 95 percent of all phone calls made to an investment firm will be answered within 20 seconds if they adopt a new phone system may be much more complex. Nonetheless, it is worthwhile to understand a few basic principles which can at least help you to make sensible estimates of probabilities.

Random Experiments and the Definition of the Probability of an Event

One way to have a consistent approach to probability calculations is to think of the event to which you want to assign a probability as a possible outcome of some experiment which is potentially repeatable over and over. Then the relative frequency with which your event occurs is its probability. These "experiments," which may be real or may possibly only happen in your mind, are called **random experiments**, because the outcome at each repetition is (usually) not certain. In this sense, picking an account at random from company records is a random experiment, as is picking an employee's name at random from among all employees, or picking an item at random from a production line. Using the idea of random experiments, we can introduce some standard abbreviations which simplify writing probabilities. They are shown in Table 4-1.

Using this notation, the definition of the probability of an event, A, is

Table 4-1. Standard Abbreviations for Writing Probabilities

Abbreviation	Concept
A (or B, C, etc.)	Some event of interest—some potential outcome or group of outcomes of the experiment.
n	Total number of possible outcomes of the experiment.
f	The number out of all the possible outcomes which correspond to (are favorable to) the event of interest.
P	"P" stands for the phrase, "The probability of."

given as $P(A) = f/n$, the ratio of the number of "favorable" outcomes to the total number of possible outcomes. In order to use this definition it must be assumed that each of the possible outcomes (which were counted to get the value of n) has the same chance to occur; they are therefore called **"equally likely"** outcomes.

According to this definition, because f cannot be more than n, the probability of an event is a fraction which can only take on values from zero to one.

Example 4-5: This is the kind of calculation that would be a small part of the design of a quality control scheme. Imagine some assembly process which produces an electronic component. The inspection routine calls for picking one piece at random from each batch of fifty; if the piece is defective then the batch is put aside for further investigation, otherwise the batch is shipped on. Suppose that, unknown to the inspector, in a batch of fifty pieces there are three defective ones. If the inspector picks one at random from this batch for testing, what is the probability that the selected one will be defective and the batch therefore rejected?

Solution:

The random experiment: picking a piece at random from the batch.

The event of interest: Let A be the event that the selected piece is defective.

The number of possible outcomes: $n = 50$, since there are 50 possible pieces to be picked, and they are all equally likely to be picked.

The number of favorable outcomes: $f = 3$ since there are 3 defective pieces in the batch.

The probability of picking a defective piece: $P(A) = f/n$

$= 3/50$

or 0.06

or 6 percent.

Interpretation: If an assembly process produces defective pieces at the rate of 3 per batch of 50, then picking one at random to inspect out of each batch will result in 6 percent of the batches being "caught" for further investigation. Knowing this, management can decide whether or not this is adequate control. For comparison, I will show later in the chapter that picking 2 at random from each batch increases the chances of catching a bad batch to about 12 percent.

Note: It is also possible to compute this probability by subtracting the probability of the complementary event from 1.

Event A:	picking a defective piece
	(There are 3 such pieces.)
Complement of A:	picking a piece which is not defective
	(There are 47 such pieces)
P(complement of A)	= 47/50
	= 0.94
	= 94 percent

Since the complement of A has a 94 percent probability, then A has a 6 percent probability. $(1 - 0.94 = 0.06)$

Conditional Probability

> **Key Concept 19**
>
> *The conditional probability of an event is the probability that it will occur given certain specific preconditions.*

Consider the following three statements:

1. The probability that an employee of the ABC corporation will eventually be fired because of drug abuse is 10 percent.
2. The probability that a *male* employee of the ABC corporation will be eventually be fired because of drug abuse is 12 percent.
3. The probability that a *female* employee of the ABC corporation will eventually be fired because of drug abuse is 3 percent.

In this scenario the population is all employees of the ABC corporation. The second two statements describe specific subgroups of this entire population. When a probability refers to a particular subgroup, it is called a **conditional probability**. A more formal way of expressing this using mathematical jargon is as follows. "*Given that an employee is male,* the probability of his eventually being fired from the ABC corporation for drug abuse is 12 percent." The expression "given that" is the mathematician's way of expressing conditional probabilities, of warning that the probability value refers to some specific group, or that certain preconditions hold.

When the probabilities of an event are different for various subgroups of the entire population, it is important to point this out by using conditional probabilities. Likewise, it is crucial in interpreting probabilities to be very clear about which group is being described. Always be alert to the possibility that a probability value mentioned for a whole population may be quite different for some special subgroup or special circumstances.

Independent Events

Suppose that on a company-wide basis, 8 percent of phone contacts result in sales. We could say then that *over all*, the probability of a phone contact resulting in a sale is 8 percent. But if there is a large difference between the percentages for the eastern and western sales regions, it may be important to report these individual probabilities. In the language of probability we say in this case that sales *depend* on the region. The probability that a phone contact results in a sale depends on the region in which the call is made.

In contrast, if the probability of a sale is the same in both regions, then we say that sales are **independent** of region. Informally, you could say that region "doesn't matter" as far as sales are concerned.

Example 4-6: Based on the table below, is it correct to say that sales response is independent of region?

	Contacts	Sales
Region A	150	12
Region B	225	18
Total	375	30

Solution: By expressing each of these as relative frequencies or probabilities, we can see that they are all equal, thus indicating that for all practical purposes region does not matter, and that sales response is independent of region.

	Contacts	Sales	Relative Frequency
Region A	150	12	12/150 = 0.08
Region B	225	18	18/225 = 0.08
Total	375	30	30/375 = 0.08

Note: The numbers in the previous example are fictitious and made up to make a point. In a "real" study, it is unlikely that two relative frequencies which are being compared will turn out to be precisely equal.

In the real study, the question would be "Are the two values *close enough* to suggest that region doesn't matter?" There is more about evaluating such differences in Chapter 6.

Which Conditional Probability Do You Want?

Many mistaken inferences are made because of careless interpretations of a conditional probability. For example, suppose you hear that 95 percent of top management positions at ABC corporation go to people already within the firm. You, being already with the firm, may conclude therefore that you have a good chance of getting a top management position. This may or may not be true, but, as is explained next, it does *not* follow from the given probability.

The situation just described says that if you look at the people who are *already* in management, then you will see that 95 percent of *them* came from within the firm. The group being described is those who are *now* in top management. Using the technical language of conditional probability, we would say: *Given* that someone is *already* in top management, then there is a 95 percent chance that he or she came from within the firm.

But since you are not in top management and you want to know *your own* chances of getting there, this 95 percent figure does not apply to you. You need a figure based on the group of *eligible* employees; that is, given that someone who works for the firm is eligible, what are his or her chances for promotion? Note that the two conditional probabilities start with different "givens"; one is starting with current employees eligible for promotion; the other with people already promoted. It is important to know which of these figures is the one you want.

Example 4-7: Last year within company ABC there were 500 people eligible for promotion to 100 new management slots. Ninety-five of them were promoted and five people were brought in from outside the firm. Assume that these ratios represent current company policy for granting promotions.

a. Find the probability that an eligible insider receives a promotion.

b. Find the probability that a person in a new management slot came from inside the company.

Solution:

a. This is a question about what happened to the eligible insiders. There were 500 such people, and 95 of them, or 19 percent, got promotions.

If this policy still holds, there is a 19 percent probability that an eligible insider will be promoted. This is the figure an employee hoping for a promotion cares about.

b. This is a question about people who already have received new management slots. There were 100 such people, and 95 of them came from within the company. If this policy continues to hold there is a 95 percent probability that a holder of a new slot came from within the company. This is the figure someone concerned about company management style might care about.

Example 4-8: You hear that 90 percent of employees fired last year had drinking problems. Does this mean that an employee with a drinking problem has a high probability of being fired?

Solution: Not necessarily; you can't tell from the given information. The given statement tells about the group of people who were fired. It does not tell about the group of people who have drinking problems. Consider a ridiculous illustration that has the same logic. A careful study of all employees fired last year showed that 99 percent drank milk as a child. You cannot deduce from this that an employee who drank milk as a child will probably end up being fired.

The main point of this section is to remind you to carefully describe the group to which a conditional probability is being applied.

Inferring Conditional Probabilities from Tables

One situation where clear probabilistic reasoning is helpful is in the presentation and interpretation of certain tables. Imagine an employee training program for 1000 people which includes the assignment of two projects, an introductory one and an advanced one.
The results of the program are first reported as follows.

On project 1, 505 of the 1000 trainees did a good job.

On project 2, 498 of the 1000 trainees did a good job.

You might be tempted to say that the training program has about a 50 percent success rate. However, a careful reading of the more detailed table below shows that the story is not that simple. In the table, each trainee is counted in the cell which corresponds to his or her particular combination of outcomes for the two projects. What is striking about this tabulation is that very few trainees did well on both projects (only 50/1000, or 5 percent).

		Project 2 Assessment		
		Good	Not good	Total
Project 1 Assessment	Good	50	455	505
	Not good	448	47	495
	Total	498	502	1000

There seems to be some disparity between how a person does on the two tests. How well a person does on one test seems related (inversely) to how he or she does on the other. As soon as you notice this dependence, you should think it appropriate to express the findings by using conditional probabilities. As an illustration we interpret the table to describe trainees' success with project 2, the advanced project.

From the table we can read:

1. Four hundred ninety-eight out of 1000 trainees did well on project 2. Therefore:
   The probability that a trainee does well on project 2 is 49.8 percent.
   When we say "a trainee," this is short for "a trainee picked at random from the entire population of trainees." This is an unconditional probability.
2. Of the 505 people who did well on project 1, 50 also scored "good" on project 2.
   Therefore:
   The probability that a person *who did well the first time* will also do well the second time is 50/505 or 9.9 percent.
   This is a conditional probability: *Given* that a person did well on project 1, the probability is 9.9 percent that he or she did well on project 2.
3. Of the 495 trainees who did poorly on the first project, 448 did well the second time.
   Therefore:
   The probability that a person *who did poorly the first time* will do well the second time is 448/495 or 90.5 percent.
   This is a conditional probability: *Given* that a person did poorly on project 1, the probability is 90.5 percent that he or she did well on project 2.

Because the success rate on project 2 is so different for these two groups of trainees, one should report the conditional probabilities separately rather than just report that 49.8 percent of all the trainees did well on project 2. The main point here is to stress that in processes that have more than one stage, tables and reports should reflect the stage structure. From a management perspective this is a most peculiar training program which deserves careful review.

Tree Diagrams and the Multiplication Rule of Probability

Key Concept 20

A tree diagram is a way of displaying sequences of events which simplifies the calculation of probabilities.

For computing probabilities in processes that have several stages you may find **tree diagrams** helpful. These are sketches on which probabilities can be recorded and combined according to a few simple rules. As long as the situation being analyzed is not too complex, the tree diagram is a handy device. As an illustration we present the information from the illustration above in a tree diagram (Figure 4-1).

Key Concept 21

The Multiplication Rule is a formal calculation procedure for finding the probability of a sequence of events.

Once the probability fractions are assigned to the various branches, the basic rule for computing the probability of the occurrence of any particular path is to multiply the fractions along the path. This is called the **Multiplication Rule of Probability** theory, which says that the probability of a sequence of events occurring is the product of a string of conditional probabilities. This is illustrated using the tree diagram in Figure 4-1.

1. What is the probability that a trainee will first do poorly on project 1 and then go on to do well on project 2? (What percentage of the trainees do poorly on project 1 and then do well on project 2?)

   This corresponds to the path from "Start" to "Not Good" to "Good."

   The probability is therefore

$$(495/1000) \times (448/495) = 448/1000 = 0.448 = 44.8\%$$

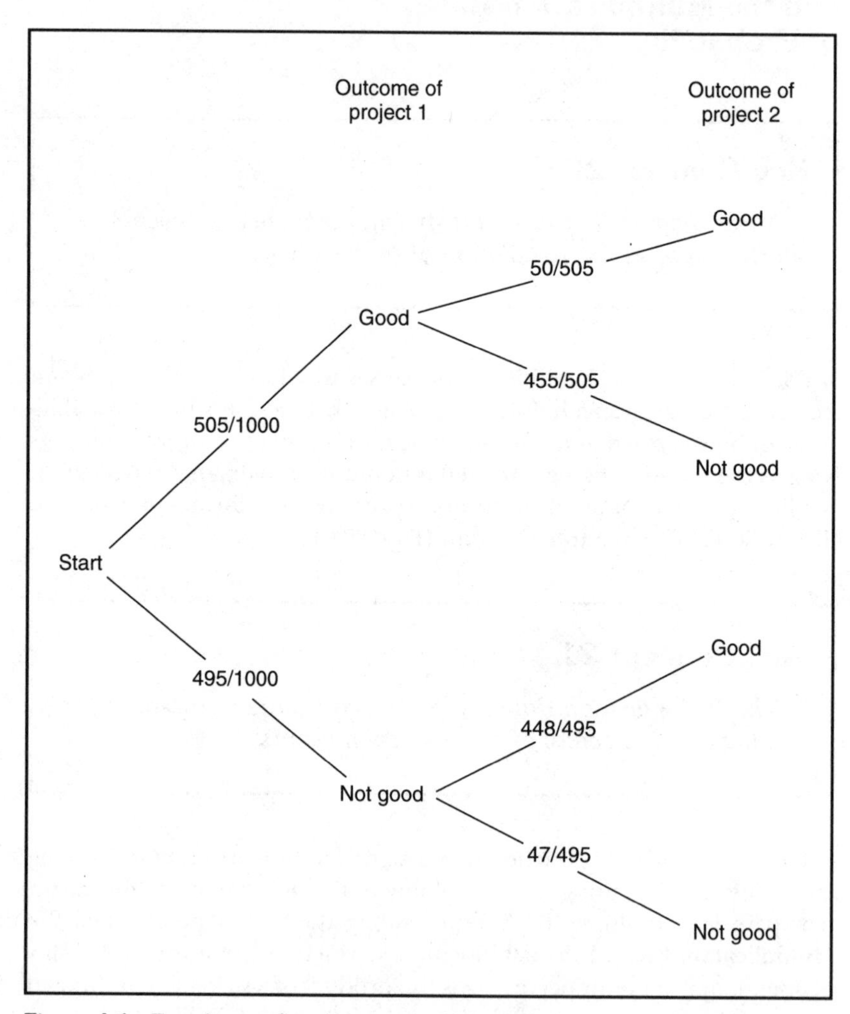

Figure 4-1. Tree diagram for table on page 82.

This reflects the fact that 448 trainees out of the original 1000 did well on project 2 after doing poorly on the first project.

2. What is the probability that a trainee will do well on project 2? (What percentage of trainees do well on project 2—regardless of performance on project 1?)

   There are two paths which end in doing well on project 2, the path from "Start" to "Good" to "Good," and the path from "Start" to "Not Good" to "Good." The solution is to apply the multiplication rule to each path and then *take the sum of the probabilities for each path.*

   Probability for path 1 (Start—Good—Good):

$$(505/1000) \times (50/505) = 50/1000 = 0.05 = 5\%$$

   Probability for path 2 (Start—Not Good—Good):

$$(495/1000)(448/495) = 448/1000 = 0.448 = 44.8\%$$

   Sum of probabilities for the two paths:

$$0.05 + 0.448 = 0.498 = 49.8\%$$

This reflects the fact that 498 of the 1000 trainees did well on the second project.

3. What is the probability that a trainee does well on project 2 even though he or she did poorly on project 1?

   In this question we are *given* the result on the first project, so we begin reading the diagram *in the middle* of the path that began "Not good." We just need to know the probability of going from there to "Good."

   Probability for path (Not good—Good):

$$448/495 = 0.905 = 90.5\%$$

This reflects the fact that 90.5 percent of the trainees who did poorly on project 1 did well on project 2.

Example 4-9: Refer to the production line inspection of Example 4-5. In that illustration the inspection scheme was to pick 1 piece at random from each batch of 50. Suppose that the decision is now made to sample 2 out of each batch of 50. How does that affect the probability of rejecting a batch which includes 3 defective pieces?

Solution: We can set up a tree diagram with two stages as shown in Figure 4-2.

This time the computations are simpler if we start with the complementary event. The event of interest is rejecting the batch, which occurs on two different paths in the diagram. The complementary event, keeping the batch, occurs on only one path.

$$
\begin{aligned}
P\,(\text{keep batch}) &= (47/50)\,(46/49) \\
&= .882 \\
&= 88.2 \text{ percent} \\
P\,(\text{reject batch}) &= 1 - .882 \\
&= .118 \\
&= 11.8 \text{ percent}
\end{aligned}
$$

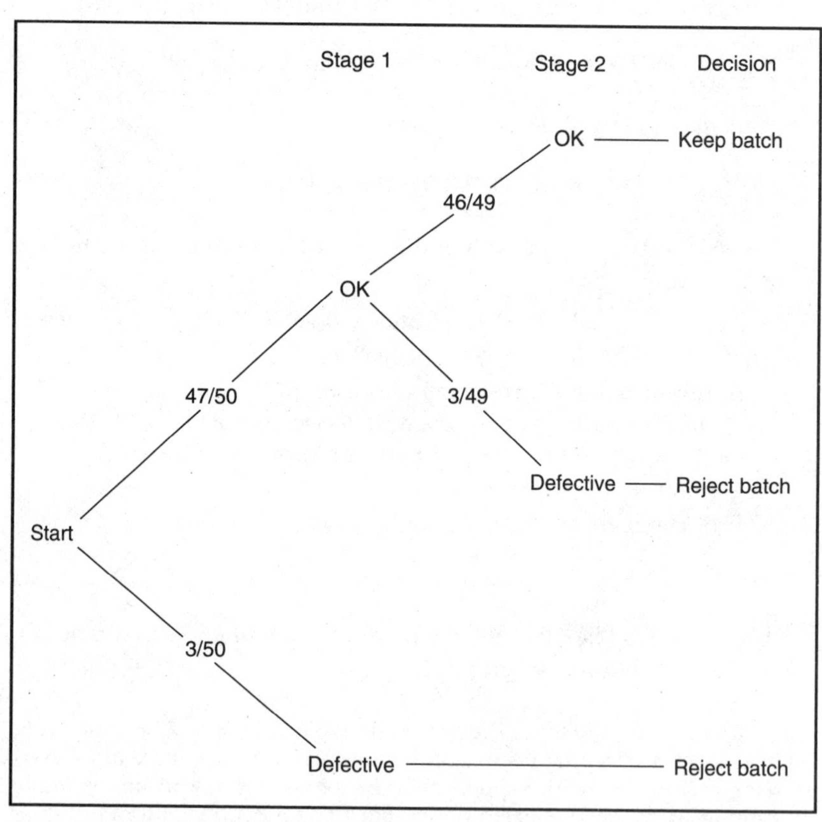

Figure 4-2. Tree diagram for Example 4-9.

Conclusion: By inspecting 2 pieces in each batch of 50 the chances of catching a batch which contains 3 defective pieces is 11.8 percent.

Note: As you increase the number of pieces inspected from each batch, a general pattern emerges as follows.

Number Picked	Probability of catching a defective piece	
1	1-47/50	= 6.0 percent
2	1-(47/50) (46/49)	= 11.8 percent
3	1-(47/50) (46/49) (45/48)	= 17.3 percent

If management's criterion was to set the number picked so that there was at least, say, a 50 percent chance of catching a "bad" batch, this pattern could be extended, and you would find that inspecting 11 pieces from each batch would be sufficient.

Expected Value

Key Concept 22

The Expected Value of a random variable is its mean value over the long run.

The **expected value** of a random variable is its long run average value. This is often of use when the issue is monetary value. If we say that the expected value of an investment is $50,000, this means that in a long series of such investments the average return is $50,000. Expected value can be used to compare risks among several options. To compute the expected value, you multiply each possible outcome by its own probability and then take the sum of all these products.

Example 4-10: A business venture with an initial investment of $25,000 has the potential to bring in $400,000 in revenues. But it is not a sure thing; there is a 20 percent probability that the venture will fail and that the $25,000 will be lost. Find the expected value of the investment.

The expected value of the venture is computed as

$$E = \text{probability of success} \times \text{value of success} + \text{probability of failure} \times \text{value of failure}$$

$$= .80(400{,}000 - 25{,}000) + .20(0 - 25{,}000)$$
$$= .80(375{,}000) + .20(-25{,}000)$$
$$= 300{,}000 - 5{,}000$$
$$= 295{,}000.$$

The expected value of the venture is $295,000.

The relative frequency interpretation of this is that if many ventures were started like this, 80 percent would make a profit of $375,000 and 20 percent would have a loss of $25,000. The average of all these +$375,000's and –$25,000's would be +$295,000.

The general formula is:

Expected Value = (probability 1 × value 1) + (probability 2 × value 2) + etc.

Be careful that the probabilities sum to 1.

Example 4-11: You run a real estate office. You need to decide whether or not to list a property for sale for the next 90 days. You estimate your up-front costs to carry the listing at $2000. If you sell the property, you will get 6 percent of the sales price as a commission. If another broker sells the property you will get 3 percent of the sales commission. The owner wants to sell the property for $200,000. What is the expected value of this deal? What is the expected value of the deal if the owner will sell for $180,000?

Solution: On the basis of past business you estimate the following probabilities:

P (no sale) = 60%.

Net loss = $2000

P (you sell it) = 20%.

Net gain = (6% of $200,000) – $2000 = $12,000 – $2000

= $10,000

P (another agent sells it) = 20%.

Net gain = (3% of $200,000) – $2000 = $6000 – $2000

= $4000

E = (probability of no sale) × (net value of no sale) +

(probability you sell it) × (net value if you sell it) +

(probability another agent sells it) × (net value if another agent sells it)

$= .60(-2000) + .20(10{,}000) + .20(4000)$

$= -1200 + 2000 + 800$

$= 1600$

The expected value is $1600.

What if the owner is willing to offer the property at $180,000? First of all, this may change the probabilities, because the property may be more likely to sell at this price. Suppose now you estimate P (no sale) = 50%, P (you sell it) = 25%, P (another agent sells it) = 25%.

$E = .50(0 - 2000) + .25(10{,}800 - 2000) + .25(5400 - 2000)$

$= -1000 + 2200 + 850$

$= 2050$

The expected value is $2050.

From your point of view, the second deal is better. Using the criterion of expected value, there is less risk associated with the second deal.

Example 4-12: You use a team of workers to assemble your product. Over time you have found that 70 percent of the time the assembly takes 4 days, 20 percent of the time it takes 5 days, and 10 percent of the time it takes 6 days. Compute the expected assembly time, and estimate how many products this team can complete in 90 days.

Solution:

$E = .70(4) + .20(5) + .10(6) = 4.4$

On average it takes 4.4 days for them to complete a project.

Therefore in 90 days they can do about $90/4.4 = 20.45$ or 20 assemblies.

Vital Statistics and Demography

Key Concept 23

Vital statistics and demography are useful in long range planning.

There is growing familiarity in the business world with the contributions of demography to defining markets and planning sales. Magazines like *American Demographics*, published by Dow Jones & Company, have rapidly growing subscriber lists. In fact, this is a good place to look for ads from companies which help design and carry out surveys for commercial customers. Another source of professional statistical help is a university or college statistics department.

Demography is the quantitative study of human populations, and until recently the word was not applied in the commercial world; historically, it was mainly concerned with broad issues of birth, death, health, and population change. But with growing business use of statistical surveys and increased sophistication of market research techniques, it is now just as likely to be mentioned in much narrower contexts—like describing recreation preferences among wealthy Californians of Japanese descent, or estimating the percentage of new parents in big cities who will use cloth diapers.

The statistical components of demography are no different from those

Social and economic characteristics of women, 18 to 44 years old, who had a child in the last year (as of June 1988)

Characteristic	30 to 44 years old	
		Women who had a child in the last year
	Number of women (1000)	Total births per 1000 women
Labor force status: in labor force	20,797	33.1
Employed	19,893	32.4
Unemployed	904	47.7
Not in labor force	7,782	76.4
Occupation of employed women:		
Managerial-professional	6,250	38.1
Technical, sales, administrative support	8,186	30.9

Figure 4-3. (Adapted from the *Statistical Abstract of the United States, 1990.*)

we have discussed in this book. Great emphasis is put on defining target populations clearly enough to be useful and on the selection of dependable samples from those populations. For certain purposes **focus groups**, small groups of potential customers, are engaged in group discussions led by trained discussants to elicit attitudes that would be hard to express by usual "complete the survey form" techniques.

The "grandaddy" of demography is the study of **vital statistics**, where "vital" refers to "life" as in "vital signs." This area of statistics refers to the reporting of birth rates, death rates, and other rates that have significant impact on the size and health of large populations. As such, it is often a crucial part of long range planning.

Vital statistics are most often reported as rates. For a rate to make sense three components must be clear: a base population, an event of interest, and a time period. This is illustrated by reading from the 1990 edition of the *Statistical Abstract of the United States.* Figure 4-3 shows part of a table in the Vital Statistics Section.

What is the interpretation of the circled number 38.1?

The label to the left tells us the number refers to women employed at the "Management-professional" level, while the label at the top says "30 to 44 years old." The population then is all women ages 30 to 44 who were employed in management-professional jobs as of June 1988. The column labeled "Number of women" shows that there were 6250 thousand (6,250,000) such women. During the year previous to that, their birth rate was 38.1 births per 1000. That is, for every 1000 women in this group, 38.1 births occurred.

The three components for this rate are:

Population: 6,250,000 women ages 30 to 44 at the management level.

Event of interest: Births

Time period: The year ending June 1988

(For this population the birth rate was 38.1 per thousand.)

Note that it is not possible to interpret this rate *without the "per thousand."* In particular, it would be a huge error to say the birth rate was 38.1 *percent,* because percent means *per hundred.*

From the birth rate you can compute the *number* of births. There were 6250 thousand women and there were 38.1 births for each 1000 women. This accounts for 6250(38.1) = 238,125 births.

A *rate* is often a good measure to use for comparing two groups on some particular characteristic because it takes into account that the groups may be of different sizes.

Example 4-13: Compare the birth rates in the table above for the two groups: "Management-professional" and "Technical, sales, administrative support."

Solution: In the management group the rate was 38.1 per thousand. For the technical group the rate was 30.9 per thousand.

This means that a woman in the management group was *more likely* to have had a birth that year than a woman in the technical group. A rate can always be interpreted as a probability; the higher the rate, the more likely the event.

Self-Check

Answers to the following exercises can be found in Appendix A.

1. On 40 of the past 50 weeks a mail order house has done more business on Monday than on Tuesday. On this basis they say that there is an 80 percent probability they will do more business next Monday than next Tuesday. This is an example of which kind of probability?
   a. empirical
   b. theoretical
   c. subjective

2. To raise money for a nonprofit community organization a company raffles off one $100 cash prize every week. Each week 200 employees, including you, play the game by paying $1 for a chance to win.
   a. On any given week what is the probability that you will win?
   b. You would expect to win this game about (once every year, once every four years, once every 200 years).
   c. Find the expected value of your gamble. Remember that if you win your net profit is $99, and if you lose your net loss is $1.

3. Which of these values are mathematically impossible for the probability of the failure of a business venture?
   a. 0
   b. 1/2
   c. 1
   d. 2
   e. –.35

4. If the odds against you are 3 to 2, what is the probability that you will succeed?
   a. 2/3
   b. 2/5
   c. 3/5
   d. 5/10

5. If the probability of success is 90 percent, what are the odds in favor of success?
   a. 90 to 1
   b. 9 to 1
   c. .90

6. In a complex product there are two critical components. The probability of failure in the field is given in the tree diagram. Use it to answer these two questions.
   a. Suppose the product malfunctions only when both components fail. What is the probability of malfunction?
   b. Suppose the product malfunctions if either one of the components fails. What is the probability of malfunction?

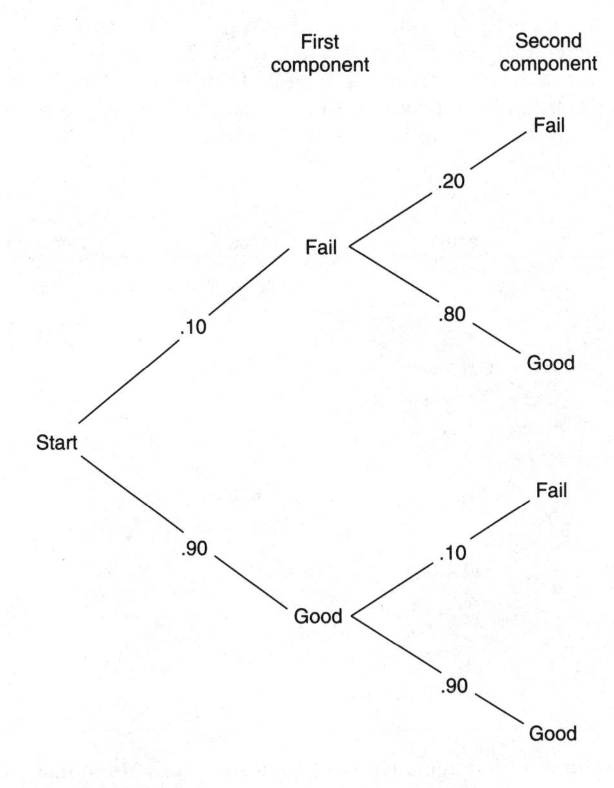

7. Let A be the event that sales increase next quarter. Which of these two events is the complement of A?
   a. Sales decrease next quarter.
   b. Sales either decrease or stay the same next quarter.

8. There are several troubleshooters who work for a business. What does this expression mean?

$$P\text{ (problem is solved given troubleshooter is Sally)} = .80.$$

9. Based on the given table:
   a. Find P (sale is made given Region A)
   b. Find P (sale is made given Region B)
   c. Find P (sale is made)
   d. Is it correct to say that sales response is independent of region?

	Contacts	Sales
Region A	200	20
Region B	600	20

10. Unknown to anyone, a production line has gone awry and is now producing 20 defective pieces in each batch of 50. The inspection scheme selects 2 pieces at random from each batch. If either one is defective, the batch is rejected and the production line is stopped. Complete the tree diagram to find the probability that the line will be stopped.

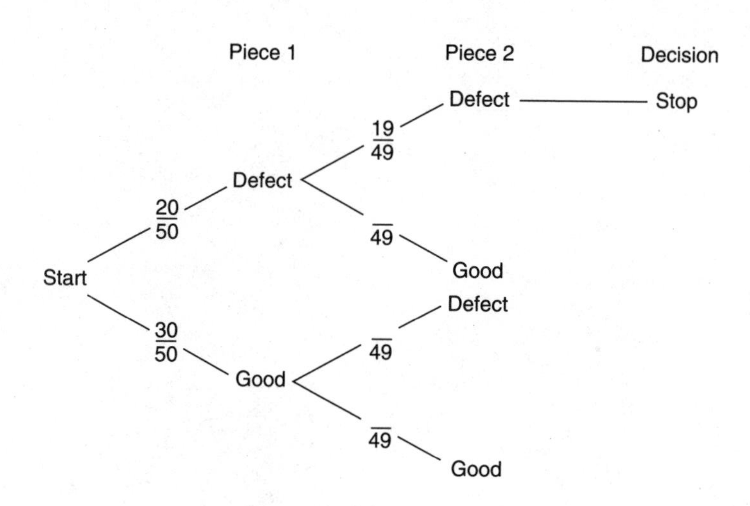

11. You are assigned to manage a type of project you have done many times before. Sixty percent of the time in the past these have taken 30 days to complete, and 40 percent of the time they have taken 45 days.
    a. What is the expected value of the time it takes you to complete one of these projects?
    b. About how long would it take to complete 10 such projects?

12. From Figure 4-3 explain the number 47.7 by answering parts a, b, and c.
    a. Describe the population; what are the characteristics of this group of people?
    b. How many women are in the population?
    c. About how many births did these women account for?

13. A statement says that in 1990 the United States birth rate was 10.1. Why is this figure not interpretable as given?

5
Estimation

Key Concepts

24. A major application of inferential statistics is the estimation of population parameters based on limited information taken from samples.
25. Because estimates are based on sample data some assessment of the potential error is necessary.
26. Random, or unbiased, sampling allows the assignment of confidence to estimates.
27. You must know the sampling distribution of a statistic to assess its reliability as an estimator.
28. For large enough samples, the sample percentage is an unbiased estimator of the corresponding population percentage, and its sampling distribution is normal.
29. For large enough samples, the sample mean is an unbiased estimator of the population mean, and its sampling distribution is normal.
30. When the sampling distribution is normal, it is straightforward to compute a confidence interval for a parameter.

The Language of Statistics

Census

Sample of size n

Sampling error

Sampling distribution of the statistic

Unbiased

Standard error

Unbiased statistic

Margin of error

Point estimate

Interval estimate

Confidence interval

Introduction

Key Concept 24

A major application of inferential statistics is the estimation of population parameters based on limited information taken from samples.

A widespread application of statistics is the estimation of characteristics of a population and the assignment of a degree of reliability to such estimates. Characteristic means something that is true about the population. The technical word for a *population* characteristic is **parameter**. You can say, therefore, that one main job of statistics is the estimation of parameters. When you can observe the *entire* population, which is called taking a **census**, then there is no question of *estimating* anything; you will know the exact value of the parameter. The converse of this is that when your study does not observe the entire population, but only a portion of it, it follows that any estimates should be accompanied by some note about the magnitude of potential error. In this chapter we describe the more commonly used techniques for making estimates, giving a little of the theory behind the techniques.

Estimation Always Involves Error

> **Key Concept 25**
>
> *Because estimates are based on sample data some assessment of the potential error is necessary.*

It should be said right off that the reality of the situation is more complicated than the theory. The elementary theory of estimation assumes that your estimate is based on what statisticians call a "random sample," which is somewhat more technical than what the nonspecialist means by that expression. To the degree that you are unable to set up a scheme to get samples that qualify as random, it is difficult to assess the reliability of your estimates. Experts should be called on when there are substantial risks involved.

To make the point, look at the warnings attached to a *New York Times*/ CBS News Poll designed to estimate voters' views on the economy (Figure 5-1). This was published November 4, 1990, but the results of similar polls are released almost every week.

This insert will be referred to later with more attention, but, for now, note that even after six paragraphs explaining their polling techniques, they still end up with this caveat: "In addition to sampling error, the practical difficulties of conducting any survey of public opinion may introduce other sources of error into the poll."

In this book we are going to assume that these "other" sources of error are not significant in order to explain the basic theory of sampling error, but in your own work you will have to bring your own experience to bear to evaluate the severity of the problem. For example, a condom manufacturer could expect a lot of "other" sources of error in a customer survey which included the question, "How often do you use a condom?" This question is so laden with social tension that it is quite likely that the responses will not be honest, and it is not obvious how to address this problem.

HOW THE POLL WAS TAKEN

The latest New York Times/CBS News Poll is based on telephone interviews conducted October 28 to 31 with 1445 adults around the United States, excluding Alaska and Hawaii.

The sample of telephone exchanges called was selected by a computer from a complete list of exchanges in the country. The exchanges were chosen to assure that each region of the country was represented in proportion to its population. For each exchange the telephone numbers were formed by random digits, thus permitting access to both listed and unlisted numbers. The numbers were then screened to limit calls to residences.

The results have been weighted to account for household size and number of residential telephone lines and to adjust for variations in the sample relating to region, race, sex, age, and education.

Some findings are reported in terms of an overall "probable electorate," which uses responses to questions dealing with voter registration, voting history, and the likelihood of voting in 1990, as a measure of the probability of particular respondents' voting on Election Day.

Theoretically, in 19 cases out of 20, the results based on such samples will differ by no more than three percentage points in either direction from what would have been obtained by seeking out all American adults. The potential sampling error for smaller subgroups is larger. For example, for blacks it is plus or minus nine percentage points.

For the question that asked respondents to rate on a scale of 1-to-10 the way things are going in the United States at the present time, the margin of sampling error was plus or minus 0.10; that is, in 19 cases out of 20, a mean rating of 4.91 would be no lower than 4.81 and no higher than 5.01. For means based on smaller subgroups the margin of sampling error is larger.

In addition to sampling error, the practical difficulties of conducting any survey of public opinion may introduce other sources of error into the poll.

Figure 5-1. *(Credit: New York Times, November 4, 1990.)*

Random Samples

> **Key Concept 26**
>
> *Random, or unbiased, sampling allows the assignment of confidence to estimates.*

We start with the definition of a (simple) random sample. The main point of the definition is that it refers to the *method* in which the sample is picked and *not* to the members of the population who actually get picked. In particular, a random sample is *not guaranteed* to be a representative sample; it could conceivably turn out to include an unusually high or low number of "unusual" members of the population. The saving grace, however, and the main reason random samples are used, is that when samples are selected at random, we can then calculate ahead of time the *probability* that they will be extreme. Remember, the goal of a survey is to determine what is *probably* true; the ability to calculate the value of this probability is based on the assumption that the data used in the calculation were collected by *random sampling.*

A sample which contains n members selected from some population is called a **sample of size** n. A sample of size n from a population is called a (simple) **random sample** if it has the same probability of being picked as every other possible sample of size n. The opposite of a random sample is a *biased* sample, which implies that some members of the population had a higher (or lower) probability of being selected than others.

The usual model for a random sample is the familiar one of drawing names out of a hat, though no one can remember the last time anyone did this for a serious business application. In fact, when local organizations pick paper raffle winners from a bag or box, it most certainly is true that some entrants have increased or decreased chances of being picked because of the way their raffle slips were folded or when they were put in the box. It's just that the risks are small and the cause is worthy, so nobody much cares about these inequities. (In amateur drawings, the last ones into the box are often the first ones to be picked because of insufficient mixing.)

The main point is that a random sample is picked *fairly* and is not "rigged" in any way. As a familiar example, think of a hand of 13 cards dealt from a deck of thoroughly shuffled cards as a simple random sample of size 13 from a population of 52 cards. Not only should you have a gut feeling that random samples provide reliable information *in the long run,* but you should also note that the mathematical theory upon which much inference is based assumes this randomness as a necessary and fundamental first step.

For simplicity, let's also assume in this book that the samples are quite small compared to the whole population. If this is not the case, then many of the calculations need to be altered or "corrected" appropriately. The basic ideas are not changed but the details of execution are.

Picking Random Samples Using a Random Number Table

In this section we describe an easy and noncontroversial way to pick a random sample from a population, with the aid of a table of random digits. Any professionally devised sampling scheme depends at some level on such a table. The use of the table is intended to remove the personal biases of the people doing the study. Most statistics books, as well as many calculators and computer programs, can provide or produce a table of digits in an apparently random order. Without going into the theory of how such lists are created, we can accept them as useful tools. Such a list is printed in Appendix C and a few lines are reproduced here. Its use is illustrated in the following example.

List of random digits—generated by computer.

8	9	8	6	3	1	8	5	8	1	8	8	4	9	1	9	6	6	9	9
7	8	5	9	5	3	7	1	7	9	6	9	3	1	9	3	2	3	3	7
9	4	3	2	4	5	6	3	4	9	4	0	3	2	1	9	2	3	6	2
7	3	1	2	1	7	5	3	4	1	7	9	0	5	3	3	3	0	0	9
0	2	1	2	8	7	2	6	6	6	5	5	4	2	1	5	5	9	1	9

Example 5-1: Use the table of random digits to pick a random sample of size 5 from a population which contains 97 members. Imagine that the population consists of the 97 clerks who work for a department of

a corporation, and that you want to interview 5 of them chosen "at random."

Step 1. Number the clerks from 1 to 97 in any convenient way, perhaps alphabetically or by length of employment or by social security number. It doesn't matter.

Step 2. Use the table to pick 5 numbers at random from 1 to 97. Because the largest number assigned to a clerk contains 2 digits, we look at the table as if it contains 2-digit numbers ranging from 00 to 99. We can start anywhere in the list and proceed straight through. For illustration, we start at the 6 in the fourth place of the first row. This gives the following set of 2-digit numbers.

63 18 58 18 84 91 96 69, etc.

Therefore we pick employees number 63, 18, 58, 84, and 91 for the interview. Any inappropriate numbers for which there are no employees are just skipped.

Sampling Distributions

Key Concept 27

You must know the sampling distribution of a statistic to assess its reliability as an estimator.

A fundamental component in the theory of estimation is the so-called "sampling distribution." We lead up to it by first describing a typical survey question and some statistical issues which arise from it. Suppose you want to estimate the percentage of people in some market who currently use your product. For this survey the "population" is *all* the people out there who are *potential* buyers. The parameter to be estimated is the percentage of the population who actually use the product.

You would start by selecting a random sample of people from this population and determining the percentage who say they use your product. Suppose 30 percent of your *sample* say they do. Note that this 30 percent is *not* the parameter, because it was not calculated using *all* the members of the population. It is called a **sample statistic**, because it was calculated using only *some* of the members of the population. But, based on this statistic, you could then infer that *about* 30 percent of the population uses your product.

Clearly, it is crucial to know the reliability of the estimate. What is meant by "about" 30 percent? How close can you expect your sample statistic to be to the parameter it is intended to estimate? If your sample estimate is 30 percent, and the population percent is really 38 percent, then you have a *sampling error* of 8 percent. What can you do to control the maximum probable sampling error? The answer to these questions is found by using the sampling distribution of the statistic, which is illustrated next.

For the sake of illustration, continue the previous example where you want to estimate the percentage of some population that currently uses your product. You decide to base your estimate on the percentage of buyers you find in a random sample of size 400 taken from the population. It is important to realize that if you were to repeat this procedure, you would get a different group of 400 people and the percentage of buyers in this second sample would almost certainly be different from the first sample statistic. Similarly, if you repeated the whole survey a third time, you would get still a different sample statistic. Imagine repeating the whole survey thousands of times, each time recording the percentage of customers in the sample. You would end up with a list of thousands of values.

The surprising and most helpful occurrence is that it is possible to predict the pattern (or distribution) of values you would get for this collection of statistics. It is this overall pattern which is called the **sampling distribution of the statistic**. In this particular case, where you repeatedly sample and record the percentage of customers in each sample, it is known that if you drew a histogram of all these values its shape would be approximately normal. In theory, the more repetitions you made the closer the shape would be to normal. The statistician sums this up by saying that the sampling distribution of the sample percentage is normal. (See Figure 5-2.)

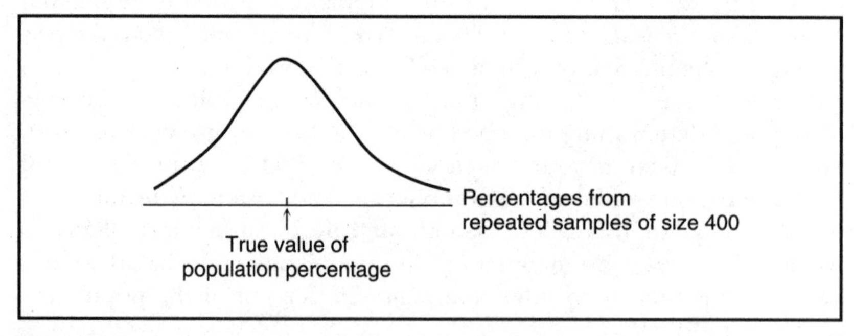

Figure 5-2. Sampling distribution of the sample percentage.

This is important because when we know the shape of the sampling distribution of a statistic, we can establish its reliability as an estimator of a parameter. Often it turns out, for example, that the *mean* of the sampling distribution is exactly the true value of the parameter which is being estimated; when this happens the statistic is called **unbiased**. Roughly speaking, this means that estimates are no more likely to come in too high or too low; in the long run the average of the estimates is right on target.

In addition, the *variability* of the sampling distribution helps determine the *probability* that a specific estimate will miss the target parameter by some given amount. Most commonly the variability of the sampling distribution is measured by its standard deviation, or **standard error**, a name which is reserved for the standard deviation of a sampling distribution.

For the most common problems of estimation it fortunately turns out that the same sampling distributions turn up over and over again. They have well-known names (often just a single letter) which are tossed off casually by statisticians. If you tell a statistician that you want to use a percentage found in a random sample of size 400 as a basis on which to estimate a population percentage, the response will be "No problem, because the sampling distribution of that statistic is normal."

Table 5-1 is a short list of some commonly used sampling distributions and some of their applications. The expertise of the practicing statistician includes knowing which one is appropriate for any given study.

This book will not go into detail on using all these sampling distributions, but will just suggest some of the issues involved. If you want more depth, you will need to consult more comprehensive texts, some of which are suggested in the bibliography. In the next few sections we show how a normal sampling distribution is used in the estimation of a population percentage and a population mean.

Table 5-1. Some Common Sampling Distributions Used for Estimating Population Parameters

Distribution	Used in the estimation of
z or Normal	population percentages and means (particularly when the samples are large)
t	population means (particularly when the samples are not large, but are taken from a normal population)
r	correlation between two variables in a population
F	variability of a population

Estimating Parameters—the Theory

The Argument from Probability

When the sampling distribution of an *unbiased* statistic is *normal,* most samples will produce estimates quite near the target parameter, and samples which produce unusually high or low estimates will be relatively rare. In particular, about 95 percent of the samples will produce estimates within two standard errors of the target parameter. (See Figure 5-3.)

But how can you make use of this theory which depends on the repeated selection of thousands of random samples when you, in fact, are only going to select one random sample? You must appeal to *probability.* You argue that though you have picked only one sample, you are *more likely* to have selected a sample which resulted in an estimate close to the parameter rather than one which resulted in an unusually high or unusually low estimate. After all, since 95 percent of the possible samples give estimates within about two standard errors of the parameter, there is a 95 percent probability that the estimate based on your one random sample is itself within two standard errors of the parameter, and only a 5 percent chance that it is more than two standard errors from the parameter. In short, *because your sample was picked at random,* there is a very high chance that it is "typical" and a very small chance that it is "rare." So basically you report your estimate and, along with it, a margin of error, which takes into account the probability that you "just by luck" selected a rare sample.

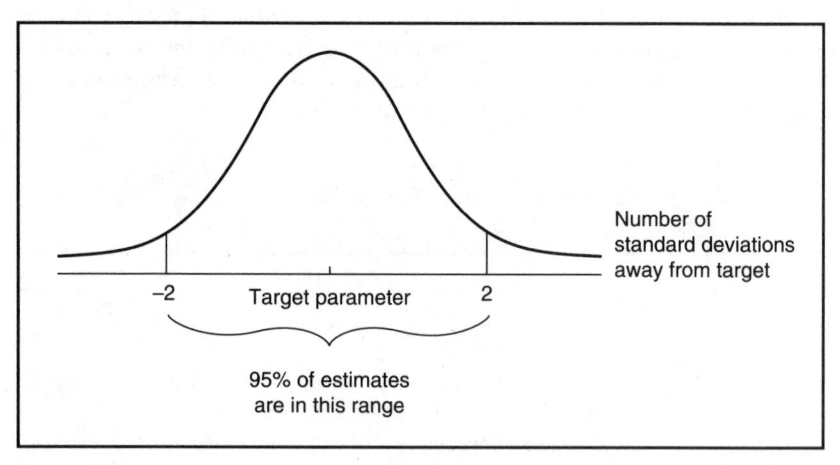

Figure 5-3

Margin of Error—Rule of Thumb for Normal Sampling Distributions

A simple rule of thumb for calculating the margin of error when the sampling distribution is normal, is to attach to any estimate of a parameter a value equal to about *two times the standard error.* The numerical value of the standard error is given by a formula worked out by mathematicians, which usually depends both on the size of the sample and the inherent variability in the population. As a consequence, you can reduce the margin of error for an estimate by selecting a larger sample from the population. In practice you must strike a balance between your desire for maximum reliability and the cost of increasing the size of the sample.

Estimating Percentages—The Computation

Key Concept 28

For large enough samples, the sample percentage is an unbiased estimator of the corresponding population percentage, and its sampling distribution is normal.

We show the computations by going through a specific example. The theory used is:

1. The sample percentage is an unbiased estimate of the population percentage.
2. When the sample size is large enough, the sample percentage has a normal sampling distribution. (As a rule of thumb, a sample is large enough if it includes at least five "yes's" and five "no's" on a "yes-no" question.)

Example 5-2: Estimate the percentage of people ages 25 to 34 in Minneapolis who listen to a certain radio station. Base your estimate on a random sample of size 400.

Population: All adults age 25 to 34 in Minneapolis.

Parameter to be estimated: The percentage that listens to a certain radio station.

Question asked: Have you listened to the radio station during the last week?

Sample: 400 people from this population picked at random.

Sample data: The number of people in the sample that answered "yes" to the question. Suppose it was 100.

This is 100/400 = 25%

Inference:

The best single estimate is that 25 percent of the market listens to the station. A single number given as an estimate is called a **point estimate**.

Margin of error:

This figure is computed by using the sample data to compute an estimate of the standard error of the sampling distribution.

The formula for the standard error of the estimate of a percentage is $\sqrt{\frac{p(1-p)}{n}}$ where p is the true value of the percentage being estimated. We use the formula by replacing p by its point estimate, .25.

$$\text{standard error} = \sqrt{\frac{(.25)(.75)}{400}} = \sqrt{\frac{.1875}{400}} = \sqrt{.0004688} = .02165$$

margin of error = 2 standard errors = 2(.02165) = 0.0433
= 4.3%
about 4%

NOTE 1: A more precise computation would use the value 1.96 instead of 2 because in a normal curve the area between $z = -1.96$ and $z = +1.96$ is slightly closer to 95 percent than the area between $z = -2$ and $z = +2$. The margin of error would have been 4.2 percent. Usually there is no practical difference.

NOTE 2: A conservative approach which results in a larger estimate of the margin of error always replaces p by .5 in the formula for the standard error. With this approach the margin of error would have been 5 percent.

Interpretation:

Although the *sample* percentage is exactly 25 percent, there is no reason to believe that the population percentage is also *exactly* 25 percent. But we can be confident that the population percentage is somewhere between 21 percent and 29 percent. A *range* of numbers like this which is intended to include the exact value of a parameter is called an **interval estimate** for that parameter.

Degree of confidence in the estimate:

This is the probability that the interval derived from the sample data does in fact include the correct value of the parameter. In this example the probability is .95 because we took the margin of error to be *two* standard errors, knowing that in a normal distribution 95 percent of the values are within *two* standard deviations of the mean. The interval from 21 percent to 29 percent is therefore called the **95 percent confidence interval** for the percentage of the population that listens to the radio station.

Reports written for researchers familiar with statistical jargon usually refer to confidence intervals, while reports in the public press usually refer to margins of error or sampling error.

Sample Size and Margin of Error

When a study is designed, the desired margin of error and level of confidence are agreed on first. Then a statistician can calculate the sample size necessary to achieve both aims. For example, to maintain a 4 percent margin of error but increase the confidence level to 99 percent in the previous example would call for a sample not of 400 people, but of almost 800 people, which could double the cost of the survey. Many practiced researchers accept the 95 percent figure as reasonable, though of course this depends on the nature of the study. One needs to consider the penalty for coming to a wrong conclusion.

What Size Sample To Use?

Algebraically, the conservative formula for margin of error given in Note 2 above for the 95 percent confidence interval can be written

$$e = 2\sqrt{\frac{(.5)(.5)}{n}} = \frac{1}{\sqrt{n}}$$

This in turn can be solved for n to get $n = 1/e^2$, which can be used as a rule of thumb for establishing the sample size necessary to get an estimate of a percentage with a given margin of error. In Table 5-2 we show some values calculated from this formula.

Table 5-2. Estimates of Population Percentages with 95 Percent Confidence; Margin of Error Related to Sample Size

Based on Formula $n = 1/e^2$

Margin of error (e)	Sample size (n)
1 percent	10,000
2 percent	2,500
3 percent	1,111
4 percent	625
5 percent	400

Case Study

New York Times Opinion Poll

We analyze the excerpt from the *New York Times* in Figure 5-1.

They claim that "In theory, in 19 cases out of 20 the results based on such samples will differ by no more than three percentage points in either direction from what would have been obtained by seeking out all American adults."

1. Why do they say the margin of error will be "no more than three percentage points"? When they refer to 19 cases out of 20, which is 19/20 or 95 percent, we know they are expressing an estimate which corresponds to a 95 percent confidence interval. In the first paragraph they indicate that the sample size was 1,445. So applying the rule of thumb formula relating sample size and margin of error we get:

$$e = \frac{1}{\sqrt{n}} = \frac{1}{\sqrt{1{,}445}} = \frac{1}{38.01} = .026 = 2.6\%$$

   This is the justification for saying that the margin of error is no more than three percentage points.

2. They say "the potential sampling error for smaller subgroups is larger." This is correct because a smaller value of n will result in a larger value of e. They say that for blacks the margin of error is 9 points. About how many blacks were interviewed? We can estimate the number of blacks in the survey by using $n = 1/e^2$.

$$n = 1/.09^2 = 1/.0081 = 123$$

   They interviewed about 123 blacks.

3. What did they do to insure randomness? The use of the computer to get random phone numbers from within exchanges is similar to our illustration using random digits. In order to insure that certain segments of the population were included, the random samples were taken among different subgroups of the population. This is called a **stratified** rather than a *simple* random sample.

Estimating the Mean of a Population

Key Concept 29

For large enough samples, the sample mean is an unbiased estimator of the population mean, and its sampling distribution is normal.

The general idea here is the same as in estimating a percentage, but the sample data now is a collection of numbers, not just "yes" or "no" responses. The theory used is:

1. The sample mean is an unbiased estimator of the population mean.
2. The sampling distribution of the mean of large samples (more than about size 30) is normal.

Key Concept 30

When the sampling distribution is normal, it is straightforward to compute a confidence interval for a parameter.

Example 5-3: Estimating the Mean Lifetime of a Product. You manufacture something with a finite lifetime and you wish to sell it with a warranty. Suppose, for example, you sell ribbons for computer printers, or ball point pens, or automobile tires. You wish to guarantee that they will last for a certain period of use. So you need to know on average how long they last.

You will test a certain number of them yourself inhouse. Since testing them also means destroying them, you cannot afford to test too many. Intuitively, you might guess that the more items you can test the more reliable your estimate will be. That is true, but how many to test also depends on how alike the items are. If the manufacturing process has such high reliability to begin with that the finished items are close to identical, a smaller sample may provide enough information.

Population: All the model X ribbons made in your plant.

Parameter to be estimated: The mean lifetime of the ribbons.

Sample: 50 model X ribbons picked at random from the production process.

Sample data: The 50 observed lifetimes, from which you would compute the mean and the standard deviation. Suppose the *sample* mean is 2.4 million impressions, and the *sample* standard deviation is 0.2 million impressions.

Inference:

The best single guess is that the mean lifetime of *all* model X ribbons is 2.4 million impressions. This is the point estimate of the *population* mean.

Margin of error:

The formula for the standard error of the sampling distribution of the sample mean is:

$$\text{standard error} = \frac{\text{standard deviation of population}}{\sqrt{n}}$$

Since the standard deviation of the population is unknown, we use the sample standard deviation as a reasonable value in its place. The working formula then is

$$\text{standard error} = \frac{\text{sample standard deviation}}{\sqrt{n}}$$

$$= \frac{0.2 \text{ million}}{\sqrt{50}}$$

$$= \frac{0.2 \text{ million}}{7.07} = 0.028 \text{ million}$$

$$\begin{aligned} \text{margin of error} &= 2 \text{ standard errors} \\ &= 2(0.028) \text{ million} \\ &= 0.056 \text{ million} \end{aligned}$$

about 56,000 impressions

Table 5-3. Margin of Error for the 95 Percent Confidence Interval Estimate of the Mean of a Population; Multipliers Based on the *T-Distribution*

Formula: $e = t \times$ Standard Error

Sample size	Margin of error
30 and over	2.00 standard errors
25	2.06
20	2.09
15	2.14
10	2.26
5	2.78

Based on this study we would have good reason to believe that mean lifetime for *all* ribbons of this model is between 2,344,000 and 2,456,000 impressions.

Degree of confidence:

Because we used two standard errors to get the margin of error, this is a 95 percent confidence interval. This is the probability that the interval from 2.344 to 2.456 million impressions does include the true mean lifetime.

Small Sample Case

The statistician uses slightly different calculations depending on whether the sample qualifies as large or small, and whether it is reasonable to assume that the population was itself approximately normal. If the sample is not large (less than about 30), but the population is close to normal, then the sampling distribution is the *t-distribution,* the graph of which resembles a slightly squashed normal curve. The main effect in the calculation of the margin of error is to increase the multiplier in the formula. See Table 5-3, which includes the *t-values* for selected sample sizes. You can see that unless the sample is really very small, you can still use a multiplier of 2 as a rule of thumb.

Self-Check

Answers to the following exercises can be found in Appendix A

1. The percentage of people in Florida who have private medical insurance is called a:
   a. parameter
   b. statistic
2. Fifty percent of the people in a sample of 2000 citizens of Florida had private medical insurance. The 50 percent figure is a:
   a. parameter
   b. statistic
3. In a sampling scheme, if every possible sample of size 40 has the same chance of being selected, then the sample chosen is said to be
   a. random
   b. biased
4. A sample with 50 values in it is called a sample of __________ 50.

5. If you use a list of random digits to pick a sample of size 5 from a population of size 870, you should think of the table as a list of (one, two, three) digit numbers.
6. Starting with the first 6 in the table on page 104, select a random sample of size 5 from a population of size 870.
7. If you repeated a survey over and over, each time asking 100 people selected at random if they thought big business was sufficiently concerned with environmental issues, then each time you would get a somewhat different percentage of who said "yes."
   a. This collection of percentages is called the __________ of the sample percentage.
   b. You would expect the histogram of this distribution to have what shape?
8. For a statistic which has a normal sampling distribution, the margin of error on the estimate of the corresponding parameter is equal to approximately __________ standard errors.
9. You wish to estimate the percentage of mutual funds which invest in tobacco products. You select 50 funds at random and find that 40 invest in tobacco products.
   a. What is the point estimate?
   b. What is the margin of error for the estimate (assuming a 95 percent confidence level)?
   c. What could you do to reduce the margin of error?
10. You wish to estimate the percentage of loans granted by a large bank that are more than two months in arrears. You want your margin of error to not exceed 3 percent. How many loans should be in your random sample?
11. You wish to estimate the mean age of people who fly first class on a certain airline's route between Los Angeles and Seattle. From a random sample of 100 such fliers, you compute a mean age of 54.8 years with standard deviation 9.8 years. What is the estimate, and what is the margin of error?
12. In a large city a survey was set up to deliberately include 40 percent black, 40 percent hispanic and 20 percent white respondents, in order to reflect their percentages in the population. Is this a stratified or a simple random sample?

6
Comparing Populations

Key Concepts

31. When populations are compared on the basis of samples taken from them, this is another form of statistical inference which is therefore susceptible to sampling error.
32. When comparing populations it is necessary to account for any inherent variability. A difference between populations is statistically significant only when it outweighs such variability.
33. Statistical significance is not the same as practical significance. The difference between two populations may be statistically significant, but of little consequence for business.
34. A type I error occurs when sample evidence suggests that populations differ, but in fact they do not. A type II error occurs when sample evidence suggests that populations do not differ, but in fact they do.
35. A hypothesis test is a formal statistical procedure for testing competing claims about populations. The details vary from study to study but the overall logic is the same.
36. A test statistic is a summarizing number computed from sample data. Its value is used to support one or the other hypothesis. For a test statistic to be useful its sampling distribution must be known.

37. A statistical study should be designed from scratch to answer specific questions. It is a wasteful approach to just collect data and then try to find out what it means.

The Language of Statistics

Statistically significant
Type I error
Type-II error
P-value
Significance level
Hypothesis test
Null hypothesis
Alternative hypothesis
Test statistic
Sampling distribution
Rejection region
Power
Contingency table
Chi-square distribution
Observed values
Expected values
P-value
Experimental design
Randomized block design
Matched pairs design
Power curve

Introduction

Key Concept 31

When populations are compared on the basis of samples taken from them, this is another form of statistical inference which is therefore susceptible to sampling error.

In Chapter 5 the main issue was the estimation of a parameter for a given population. An equally common application of statistics involves comparisons among two or more populations. To illustrate this, consider the following three statements, each of which might have been made as the result of a statistical study.

1. Of the three magazines we might advertise in, the readers of *The New Yorker* have the highest mean household income.
2. Young employees in our corporation have a higher rate of absenteeism than older ones.
3. The new version of the product lasts longer than the old one.

Each of these examples implies that some comparison was made among relevant populations, and that a difference was found. In the context of this chapter, it is important to remember that the comparisons were based on data from *samples,* and that in each case the entire population was not available. Thus, these statements are inferences made on the basis of incomplete information.

As with estimation, when you do have complete information for the entire populations, there is no inference is to be made. What you see is what there is. For instance, suppose your company has two offices and you wish to compare the sales generated by them. If that is the whole point of your study, and if you have all the sales information, then there is no larger population for which an inference is to be made. You must just decide if the difference you find between the two offices is of any *practical* importance. This is a business judgment, not a matter of statistical inference. It is only when you generalize to people who were not represented in the original data that you are making an inference, and at that point the possibility of error becomes a factor to be considered.

Accounting for Inherent Variability—Statistical Significance

Key Concept 32

When comparing populations it is necessary to account for any inherent variability. A difference between populations is statistically significant only when it outweighs such variability.

How does the proper use of statistical procedure help when you make comparisons based on samples? Mainly by indicating when observed differences are more likely to be due to the inherent variability within the populations or more likely due to some difference between the populations beyond inherent variability. Perhaps this is best illustrated by an example (deliberately too small, but manageable).

Example 6-1: Suppose you routinely give some kind of aptitude test to prospective employees. The question arises whether young people score higher than old people. You pull records randomly for nine people from each group. You see that the mean for the nine young people is 86 and the mean for the nine old people is 82.

Should you conclude from this that, *in general,* young people score higher on this test than old people do? Would you be secure, for example, in inferring that for the entire populations of old and young job applicants the mean score for all young people would be higher than the mean score for all old job applicants? An important point is that any inference from this data must take into account the inherent variability in the test scores. Shown below are two sets of samples of test scores which would lead to different conclusions.

Scores on aptitude test given to prospective employees.

Data set 1.

Young people:	87, 85, 86, 87, 86, 85, 87, 85, 86 Mean = 86
Old people:	81, 82, 83, 83, 81, 82, 82, 81, 83 Mean = 82

Data set 2.

Young people:	78, 94, 81, 91, 76, 96, 78, 94, 86 Mean = 86
Old people:	76, 88, 68, 96, 72, 92, 70, 94, 82 Mean = 82

First, note that although the means are the same in both sets of data, the two sets do not give the same impression. From data set 1 it seems clear that in general the young people score higher. All the young people score about the same and they all score higher than the old people. The observed differences between the means, 86 and 82, is "real"—it represents the fact that, in general, young people score about four points higher than old people do.

Data set 2 tells another story. There is so much variability within each group, that the four point difference between the means does not represent any consistent difference between young and old. We would be on much shakier ground deciding on the basis of this data that in general the younger score higher than the old.

When these data are submitted to the official calculations of a formal statistical test, the results would be as follows. Data set 1 would lead to a finding of a **statistically significant** difference. Data set 2, would yield "no significant difference." In both cases the observed difference between the means is four points, but the statistical computations would take into account the inherent variability within the groups, and distinguish between the two cases. When a difference is called statistically significant, it means that the difference between the two sample statistics outweighs the inherent variability in the two samples. It implies that the samples differed sufficiently for us to be confident that if both entire populations were to be surveyed, the two parameters being compared (in this case, the two population means) would be found to be unequal. The calculations reflect the fact that the more variability we see in a sample, the more variability there probably is in the population from which it came—and the more difficult it may be to make inferences with much certainty from the sample to the population.

Statistical Significance Versus Practical Importance

Key Concept 33

Statistical significance is not the same as practical significance. The difference between two populations may be statistically significant, but of little consequence for business.

Nevertheless, still referring to the data of Example 6-1, it is important for you as the interpreter of these statistics to decide—even in the case of data set 1—whether four points has any practical significance. It is still possible that the wisest response even to a statistically significant difference may be "who cares?" A finding that is statistically significant may be totally trivial from a business point of view. It is unfortunate that reports on statistics often just say "significant," thereby creating the impression that something important has been discovered.

The Unavoidable Problem of Error

Key Concept 34

A type I error occurs when sample evidence suggests that populations differ, but in fact they do not. A type II error occurs when sample evidence suggests that populations do not differ, but in fact they do.

Because your conclusion will be based on sample data, there remains the possibility that your data—just by blind chance—is not representative of the populations. In one of your samples you may have gotten a preponderance of high scores and in the other a preponderance of low scores. Again we present a made-up exaggerated example to illustrate the point.

Example 6-2: Suppose that you want to compare some component of car radios made by two different suppliers. Both suppliers have made many thousands of these components, so the populations are quite large. The data consists of ratings for the components on a special scale which indicates how well they perform under severe vibration.

Population 1: All the ratings for brand A.

Consists of many thousands of values. Let's pretend that the ratings are equally distributed as 7s, 8s, 9s, 10s, and 11s. There are thousands of each in the population.

Population 2: All the ratings for brand B.

Identical to population 1. There is really no difference between the components supplied by the two manufacturers.

Sample 1.

A sample of 50 components picked at random from the weekly output of factory A. Just by pure chance the sample contains components which all have ratings of 10 or 11.

Sample 2.

A sample of 50 components picked at random from the weekly output of factory B. Just by pure chance the sample contains components which all have ratings of 7 or 8.

The scenario in Example 6-2 *could* happen, though it is not very likely. All you would see is the 50 components from each sample. You would have no idea that they were untypical of their populations. You would decide that one brand is better than the other. And you would be wrong. The formal name for this kind of wrong conclusion is **type I error**. It occurs when you conclude that a difference exists between populations when there really is none.

In principle, the possibility of this scenario exists every time you make an inference based on samples. It comes with the territory. What can you do to decrease the chances of this happening?

1. Take the largest samples you can afford. This reduces the likelihood that you only get "weird" or untypical members of the population. Larger samples tend to be more representative of the population.

2. Make sure there are no built-in biases to the *scheme* you use to choose the sample. This reduces the chance that there is something about the *way* you are picking the sample that favors untypical members or members that are all alike. This is especially a problem for so-called "self-selected" samples, like people who voluntarily return unsolicited mail surveys.
3. Do not declare a difference proved until you are convinced that the probability of a type I error is so small that you are comfortable taking the risk. This criterion can be dealt with in the statistical analysis.

We have said that when sample data *mistakenly* lead to declaring a difference which is not really there, this is a type I error. The flip side of the coin is called a **type II error**. This occurs when the sample data mistakenly lead you to declare there is no difference when really there is. There'll be more about this possibility at the end of the chapter. For now let's concentrate on the type I error, because in some ways it is easier to control.

P-value and Significance Level

Under the assumption that samples are chosen randomly, the rules of probability allow you to compute an upper limit to the probability that your data have led to a type I error. This is called the **p-value** for the study. The smaller the p-value, the less likely your finding of a difference is in error. Analyses done by commercial statistical software routinely include p-values, so that the user can decide if the error likelihood is sufficiently small.

To avoid after-the-fact arguing about what is and what is not a "small" p-value, many studies are designed with a decision rule in advance of collecting the data. It might be agreed in advance, for instance, that only if the p-value turns out to be less than 5 percent will any difference be considered statistically significant. A study which is set up like this is said to be done "at the 5 percent **significance level**." When the p-value for a study turns out to be less than five percent, most researchers say they are "confident" that the apparent difference is real and they are willing to take action based on this difference. If the p-value is larger than five percent they say that the difference is not statistically significant.

It is interesting to note that the mathematician James Bernoulli, who was one of the first to investigate the problem of inference error (300 years ago), proposed as his own rule that an inference was what he called "morally certain" only when the probability of error was less than one-tenth of one percent. This degree of certainty requires immense samples in most circumstances, and current usage is much more likely to settle for an error probability in the one to five percent range.

Hypothesis Tests for Comparing Populations

Key Concept 35

A hypothesis test is a formal statistical procedure for testing competing claims about populations. The details vary from study to study but the overall logic is the same.

The formal approach to comparing populations is called **hypothesis testing**. The technical details of a hypothesis test depend on the specific comparison being made, the number of populations involved, and the type of data being recorded. First, I'll give an overview of what's involved, followed by several specific examples.

The following steps are part of any hypothesis test.

Step 1. A pair of opposing and mutually contradictory hypotheses are formulated. A **hypothesis** is just a statement about population parameters that may or may not be true. For studies which compare populations, one hypothesis claims that the populations are alike while the other says they are different. The hypothesis which claims that the populations are *alike* is called the **null hypothesis**. The other one is usually called the **alternative hypothesis**.

Key Concept 36

A test statistic is a summarizing number computed from sample data. Its value is used to support one or the other hypothesis. For a test statistic to be useful its sampling distribution must be known.

Step 2. Decide which statistic should be computed from the sample data in order to best contrast the two hypotheses. This is called the **test**

Table 6-1. Some Common Types of Hypothesis Tests and Associated Test Statistics

Purpose	Sampling distribution of the test statistic
Compare 2 means	Normal or t
Compare 2 percentages	Normal
Compare 3 or more means	F
Compare 2 or more percentages	Chi-square

statistic. For example, if you were comparing the *means* of two populations, you would compute the means of two samples and base your test statistic on the difference between these two means. The test statistic will be useful only if you have a way of assigning probabilities to its various possible values under the assumption that the null hypothesis is true. In other words, you must know its *sampling distribution*. In many circumstances the appropriate test statistic has a well-known probability distribution. Table 6-1 lists some of these, and we will illustrate some in our examples.

Step 3. Because the sampling distribution of the test statistic is known, probabilities can be computed for all the possible values of the statistic. You can then mark off those values that are unlikely when the null hypothesis is true. This set of values is called the **rejection region** for the test, because if the test statistic turns out to have one of these values, the null hypothesis is not likely to be true. The rejection region consists of all those values of the test statistic which have small p-values. This logic is summarized in Figure 6-1.

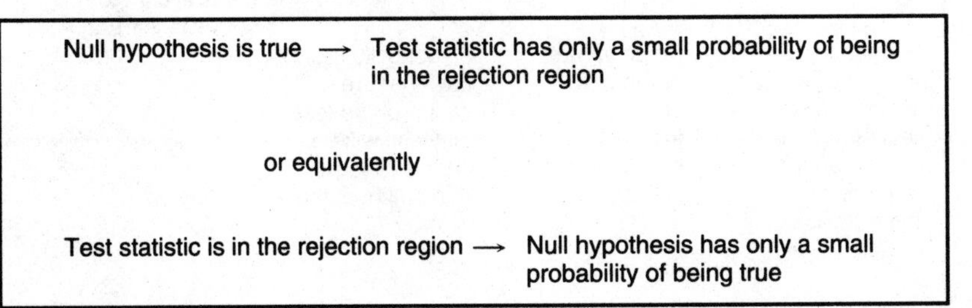

Figure 6-1. Relationship between null hypothesis and rejection region.

Step 4. Using the data, compute the test statistic and see if its value is in the rejection region. If so, the difference between the samples is statistically significant. Otherwise, it is not.

The Logic of a Hypothesis Test

The statistician's approach pretends for the sake of argument that the populations are alike. This belief will be changed only when the evidence against it is strong enough to be statistically significant. It's a little like a justice system which says innocent until proven guilty. In most studies the person doing the research, in fact, believes that the populations are *not* all alike and is hoping to garner supportive evidence. It is usually this belief that instigated the study in the first place. "Rejecting the null hypothesis" is often the real goal of a study. By "strong enough" evidence, therefore, we mean that the value of the test statistic falls in the rejection region.

Figure 6-2 shows the logic in a picture. This sketch shows what a rejec-

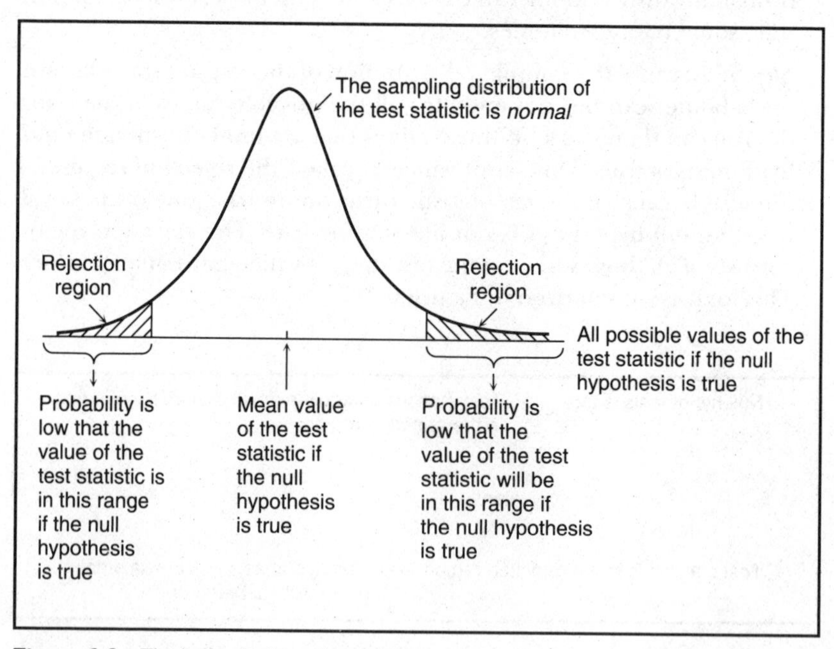

Figure 6-2. The logic of a hypothesis test.

tion region might look like in the case where the sampling distribution of the test statistic is the normal distribution. For this illustration the total rejection region is composed of two parts, one for exceptionally high values of the test statistic, and one for exceptionally low values. A test set up like this is often called a two-tailed hypothesis test.

What Happens When You Can't Reject the Null Hypothesis?

If the evidence in your study fails to confirm your beliefs that the populations differ, there are a few things you can do:

1. You can abandon your belief. Perhaps you were wrong after all. It just appeared to you that the populations differed but the evidence shows otherwise. In any case, the difference is "not statistically significant."
2. You can maintain your belief in the face of the evidence and continue to proclaim the difference between the populations, but you will have no good support, and so, for example, it would be risky to invest money based on the reality of this difference.
3. You can design a more powerful study, one that is more sensitive to any differences between the populations. Sometimes this is as simple as just continuing the study to get larger samples. Or it may involve more sophisticated ways of matching members of one population to those of the other. When statisticians refer to the **power** of a study, they mean its potential to detect differences between populations. One goal of a good study design is to get the maximum power for the time and money available.

Illustrations of Hypothesis Tests for Comparing Populations

Key Concept 37

A statistical study should be designed from scratch to answer specific questions. It is a wasteful approach to just collect data and then try to find out what it means.

The rest of this chapter is devoted to specific examples of hypothesis tests for comparing populations. There are too many different possible studies to include every possible type of test here. That is material for a more advanced text on hypothesis testing and experimental design. We will only describe a few very common situations, placing the details of the calculations in Appendix D. The main point here is to point out that though the technical details differ, the overall approach does not.

Example 6-3: A Comparison of Several Percentages. A study is set up to see if a product is selling equally well among the four ethnic groups which make up large markets in a metropolitan area. The sampling was done randomly from economically comparable populations, perhaps by picking people at random in some large shopping center, so that there are not gross differences in buying habits. Let us assume that each person was asked: "Have you used our product during the last month?"

Responses could be classified like this.

	White	Black	Asian	Hispanic
Used the product				
Did not use it				

A table like this where each cell contains the number of responses in the appropriate cross-classification is called a **contingency table**. Computing the percentages in the various categories and doing other summaries is often called **cross-tabulation** or cross-tabs for short. Most commercial software will do this for you.

First we just outline the steps of a formal hypothesis test for this data; then we work out an example with numbers.

1. *The hypotheses.*
   Null Hypothesis: All four populations are alike. The same proportion of people in each population have used the product during the last month.
   Alternative: They are not all alike.
2. *The test statistic—chi-square.*
   For comparing percentages in contingency table analysis, the test statistic is a number computed from the sample data which has a sampling distribution called the **chi-square distribution**. This statistic summarizes the difference between the data recorded in the cells of the table (the so-called *observed* values) and the numbers that *would* have been there if the data had come out *exactly* according to the null hypothesis, (the so-called *expected values*). The formula for the chi-square test statistic is given in Appendix D.

3. *Rejection region.*
   Because the sampling distribution of the chi-square statistic is known, probabilities can be computed for all the possible values of the statistic. We can then note those values that are unlikely when the null hypothesis is true (the ones with small p-values). This set of values makes up the rejection region for the test. A sketch of a typical chi-square distribution with shaded rejection region is shown in Figure 6-3.
4. If the value for the test statistic falls in the rejection region, the null hypothesis is probably not true. This means that there is strong evidence that at least two of the ethnic groups disagree on this product. If the value for the statistic is not in the rejection region, the proportions of buyers in the different groups are not different enough to warrant any special action.

Example 6-4: With Numbers. Does the following table support the hypothesis that the same percentage of people use the product in all four ethnic groups?

	White	Black	Asian	Hispanic
Used the product	24	60	36	18
Did not use it	80	100	60	62

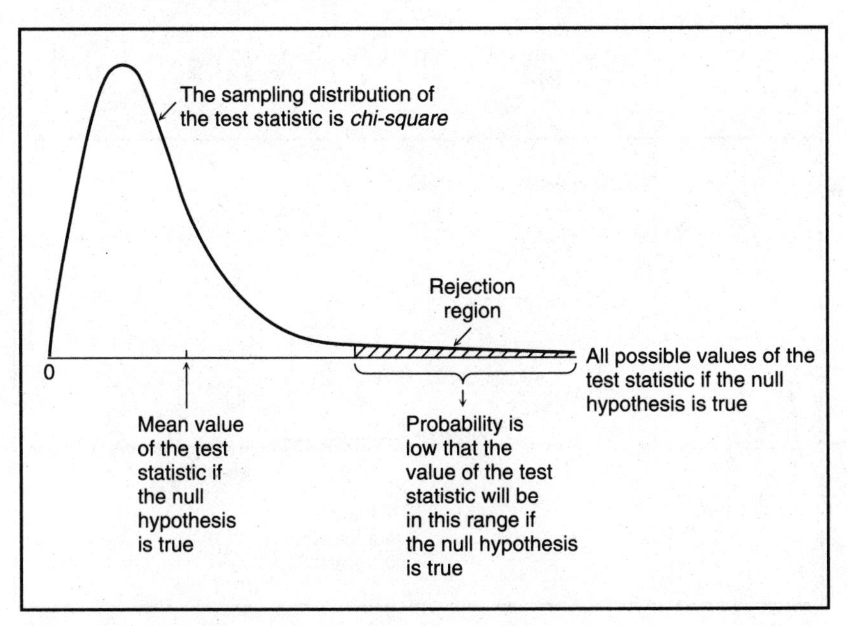

Figure 6-3. Chi-square sampling distribution.

1. The hypotheses are as stated above.
2. The test statistic has a chi-square sampling distribution. The study is done at the 5 percent significance level. By referring to a table for the chi-square test statistic or using commercial software, the rejection region is found to consist of all values greater than 7.81. The sketch is shown in Figure 6-4.
3. By doing the computation using the formula given in Appendix D, the *value of the test statistic* is found to be 10.715. This is in the rejection region. If the calculation were done by a commercial software package, it would also give the *p-value for this statistic* as 0.0062. The p-value being less than 0.05 is equivalent to the test statistic being in the rejection region, because the study is set up at the five percent significance level. (The p-value is calculated by measuring the proportion of area under the curve which corresponds to values more extreme than 10.715.)
4. Inference: Because the value of the test statistic falls in the rejection region, the null hypothesis is probably *not* true. This means that the counts in the contingency table are *not* likely to have come from four populations with the same percentage of people who use the product. Evidently, the four ethnic groups are not all alike. There *is* a statistically significant difference among at least two of the ethnic groups.

   In the language of conditional probability from Chapter 4, the two variables, "product usage" and "ethnic group," are statistically dependent.
5. Reporting the findings. The report should reflect that the original

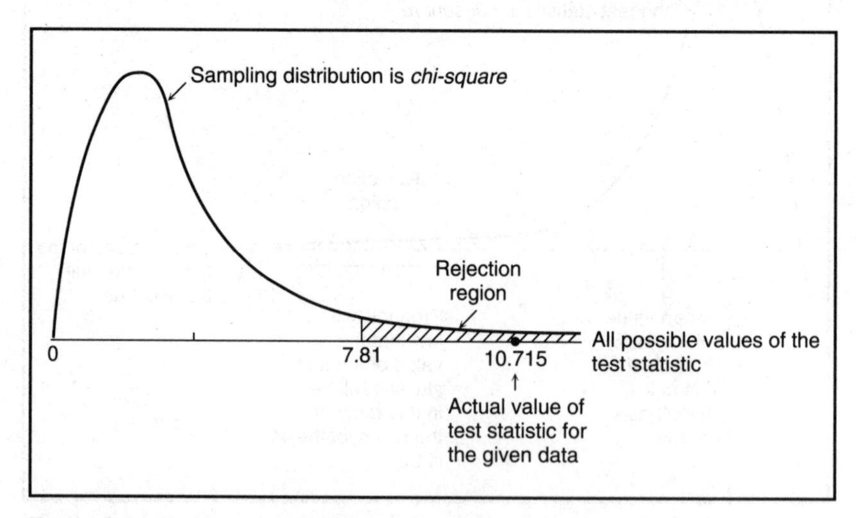

Figure 6-4. Chi-square sampling distribution for Example 6-4.

study was framed in terms of percentages. Here is one brief way to summarize the findings.

Group	Number interviewed	Percentage who said they used the product
White	104	23.1
Black	160	37.5
Asian	96	37.5
Hispanic	80	22.5

Statistical analysis by a chi-square test shows statistically significant differences among these groups. You can see that the Black and Asian groups *in the sample* have higher percentages of users. The statistical analysis indicates that this finding is strong enough to conclude that *this difference holds for the populations* they represent.

A note on the design of this study: In setting up a study like this there are several options for collecting the data. At this point I will mention three possible plans. The differences among them do not matter in the general discussion of hypothesis testing, but in a specific application of your own you or a professional statistician might prefer one to the others.

Plan 1. A simple random sample. You decide ahead of time how many people you will select at random in the shopping center. No particular effort is made to get prescribed numbers in each ethnic group. This is the design that was illustrated. If the whole point of the study is to compare the four percentages, then this is a good design.

Plan 2. Choose samples which are proportional in size to their respective populations. This would be called stratified random sampling, and is important if you want to give more weight to the larger groups. This might be the case if your study was also designed to address other questions in addition to the one we are discussing. In general, you don't want to design a study to try to answer too many questions at once. The design requirements of one question may be in conflict with those of another. The result may require huge samples and complicated interpretation of the statistics. Whenever possible, keep it simple.

Plan 3. Choose random samples of equal size. This method insures that all groups get equal weight, and often has the most statistical power. This is fine for the main question of this study, but it makes the study useless for certain other estimates, such as the percentage of the shoppers at this center who are Hispanic.

The main point to be taken here is basic. *Know what you are trying to find out* ***before*** *you collect any data.* The study should be *designed* with particular questions in mind. Otherwise you could end up spending a lot of time and money collecting useless information.

Example 6-5: Comparison of Two Means. Many studies include a comparison of means. Here is one that could be part of a study to compare two manufacturing techniques. Suppose two methods are available for assembling an appliance. You wish to decide if there is any important time difference between the two methods. It seems reasonable to do this by comparing the means of the times it takes to assemble the product by each method. You would do a trial of both methods and summarize the data like this.

	Number assembled	Mean time to assemble	Standard deviation
Method 1:			
Method 2:			

Assembling several appliances by a particular method gives a sample from the population of *all* the appliances that *will be* assembled this way. You would want to insure that this is a not a biased sample; for example, you probably would not want to begin sampling until the assembly team knew how to do the job. The first few assemblies might take an unusually long time and would not be appropriate to your main question.

We outline the steps of an appropriate hypothesis test.

1. *The hypotheses.*
   Null hypothesis: The mean assembly time for both methods is the same.
   Alternative: They are not the same.
2. *The test statistic.*
   For comparing the means of two populations, the test statistic is a number based on the difference of the two sample means and the standard deviations of the two samples. The formula is given in Appendix D. This test statistic has a *normal sampling distribution* if the two samples are large enough, or a *t-distribution* if the samples are smaller but taken from normal populations. For comparisons of more than two means, the test statistic has an F distribution, and the statistical procedure is called **analysis of variance**.
3. *Inference*
   If we get a value for the test statistic that is in the rejection region, we will decide that probably the null hypothesis is not true. We will have good evidence that one assembly method takes longer on average than the other. On the other hand, if we get a value for the statistic that is *not* in the rejection region, we will decide that the time difference between the two methods, as revealed in this study, is not big enough to warrant any special action on our part.

Example 6-6: With Numbers. Suppose the data are as follows.

	Number assembled	Mean time to assemble	Standard deviation
Method 1:	20	4.48 hrs	0.40 hr
Method 2:	20	4.43 hrs	0.43 hr
Difference:		0.05 hr	

1. The hypotheses are as stated above.
2. Because the samples are smaller than size 30, and it is reasonable to believe that assembly times are approximately normally distributed, *the sampling distribution of the test statistic is a t-distribution.* From a table for the t-distribution it is found that the rejection region consists of all values larger in absolute value than 2.024. See Figure 6-5.
3. The test statistic is computed using the data in the table (formula given in Appendix D), and its computed value is found to be 0.36. It has a p-value of 0.72.
4. Inference. Because the test statistic is *not* in the rejection region (or, equivalently, because the p-value is not small) we *cannot* say there is a statistically significant difference between the mean assembly times for these two methods. There is too much inherent variability to be sure at this point.

Example 6-7: Comparison of Standard Deviations. In some applications the issue of variability is more important than the issue of averages. For instance, in the experiment described in Example 6-4 you may be concerned that assembly time is more *consistent* with one method than the other. A hypothesis test can be set up to compare the variability of the two samples. The test statistic is the ratio of the squares

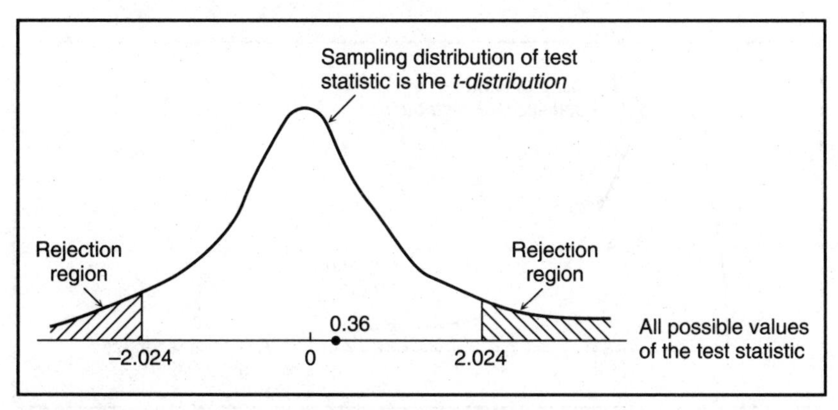

Figure 6-5. The t sampling distribution for Example 6-6.

of the two standard deviations. This statistic is used because it has a known sampling distribution, called the F distribution (named in honor of one of the founders of modern statistics, Sir Ronald Fisher).

1. *The hypotheses.*
   Null hypothesis: The two populations have the same standard deviation.
   Alternative: The standard deviations are not equal.
2. The test statistic will have an F distribution. According to tables for the F-distribution, at the .05 significance level the *rejection region* will consist of all values greater than 2.53, as illustrated in Figure 6-6.
3. For the data in Example 6-4, the value of the *test statistic* is $.43^2/.40^2 = 1.2$, which does not place it in the rejection region.
4. *Inference:* The evidence is not strong enough to reject the null hypothesis. There is no statistically significant difference between the two standard deviations. The two assembly methods are about equally consistent.

Example 6-8: Comparison of Two Means: a Matched Pairs Design. This final illustration points out that some types of comparison studies are more powerful than others. Deciding the best way to set up a study so that the data are collected and analyzed most efficiently is the concern of the field of statistics called **experimental design**. A common device for achieving more power in comparison studies is **matching**, which we describe next.

Suppose your company makes commercial dog food, and part of the research is to adjust the formula for the ingredients to get maximum weight gain in puppies. Suppose at some point you need to compare two versions of the food. You will randomly assign some puppies to one diet

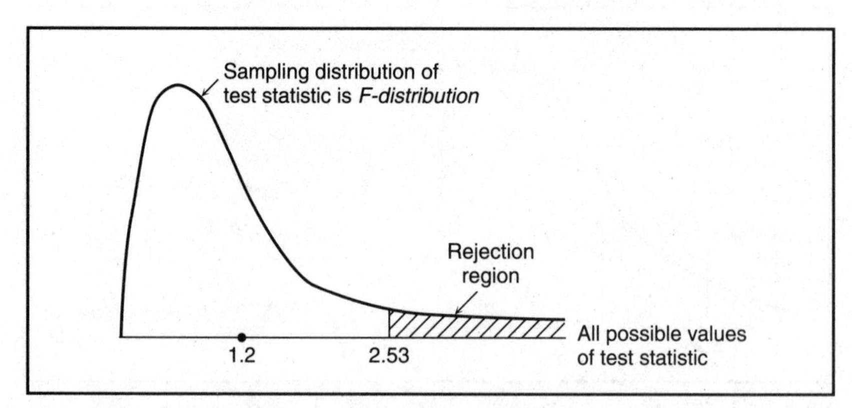

Figure 6-6. The F sampling distribution for Example 6-7.

while the rest get the other diet. The effects of the diets are obscured somewhat by any inherent differences among the dogs. Just imagine that diet 1 was given to randomly picked St. Bernards and diet 2 was given to randomly picked Chihuahuas. This is a crude example of poor experimental design. Of course the dogs on diet 1 would show more weight gain, but it wouldn't have much to do with the diet; they are just a breed that naturally grows faster.

Good design would minimize this problem right from the beginning by trying to insure that each dog in group 1 is compared to a dog in group 2 that is as much like it as possible. That way any difference in weight gain is likely to be due to diet and not some genetic difference between the dogs. Ideally, the two dogs in each pair would be clones (in fact, this is done with mice in medical research). But next best would be to use dogs from the same litter. In any case the idea is that an experiment which compares groups will be more sensitive when matching is possible. This may mean that you can use smaller samples, thereby saving money and time collecting the data.

A study that uses matching to compare treatments is called a **randomized block design**, a block being the abstract equivalent of a litter—a group of subjects that are as much alike as possible, who are then randomly assigned to the different "treatments" being compared. In the particular case where only *two* treatments are being compared it is called a **matched pairs design**, a pair being the smallest possible block. When no attempt is made to match, but the two samples are just picked randomly, the design is called **independent samples.**

Here's how the data might look for the dog food study if there were five pairs. In each pair the two dogs have been matched so that they are genetically similar.

	Weight gain on		
Pair number	diet A	diet B	Difference
1	1.8	1.4	0.4
2	2.6	2.1	0.5
3	0.7	0.5	0.2
4	1.5	1.2	0.3
5	2.0	1.6	0.4

Mean gain on diet A = 1.72 pounds
Mean gain on diet B = 1.36 pounds
Mean difference = 0.36 pounds

When this data is analyzed *taking the matching into account,* the difference in weight gain is found to be statistically significant. The computations "know" that in each pair the dog on diet A gained more weight. If the data are analyzed ignoring this, by just comparing the two mean weight gains, then the difference between the diets would not be statistically significant because the variability among the dogs on each diet would obscure the consistent difference in favor of A.

Here is a brief outline of the steps of the hypothesis test.

1. *The hypotheses.*
   Null Hypothesis: Weight gain is the same on both diets.
   Alternative: Weight gain is not the same on both diets.
2. The test statistic in the matched pairs analysis is based on the column of differences, and its sampling distribution is known to be the t-distribution. At the .05 significance level, the rejection region, according to tables for the t-distribution, consists of all values greater in magnitude than 2.78. This is shown in Figure 6-7.
3. For the given data, the test statistic turns out to equal 7.06, which is in the rejection region. The formula for the calculation is in Appendix D.
4. Inference. We reject the null hypothesis. The evidence is strong enough to indicate that in general the mean weight gain is greater on diet A.

More on Power and Error

A few issues that were touched on earlier in the chapter are worth further comment. When you choose the significance level for a study you are setting the probability of a type I error at that value. For instance, if you do an analysis at the 5 percent significance level and the samples lead you to reject the null hypothesis, then there is a 5 percent probability that you have committed a type I error. You control the probability of a type I error by your choice of significance level; it is a mathematical

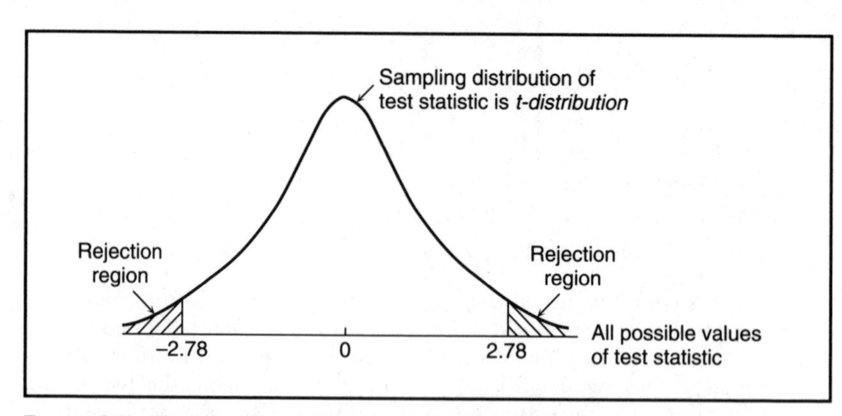

Figure 6-7. The t sampling distribution for Example 6-8.

procedure. If, for example, you are unwilling to risk a 5 percent chance of a type I error, you can make the rejection region smaller. Figure 6-8 may clarify the situation.

When you read that a study *has* found a difference between populations, and you ask "What are the chances the study is mistaken?", this is the same as asking "What was the size of the rejection region that was used in this analysis?" In contrast, a study which finds *no difference* is open to the possibility of a type II error. This error occurs when there really is a difference in the populations but the samples fail to show it. The likelihood of this error depends on *how big* a difference there really is. The bigger the population difference really is, the less likely the error will occur. As a consequence there is not one simple answer to the question "What is the probability that a type II error has occurred?"

A statistician can prepare a chart showing the probability that samples of a given size will lead to a type II error *for various possible population differences.* Such a chart (called a **power-curve**) serves as a guideline to the potential effectiveness of the study and is useful in determining how large the samples should be if you are to have a reasonable chance of establishing a difference of any given magnitude. Occasionally, for instance, such a chart shows that it would be pointless to even conduct a study as planned because it has very little chance of establishing the suspected difference between two populations. The statistician would say the design has too little power for its intended purpose and would probably suggest at least increasing the proposed sample sizes.

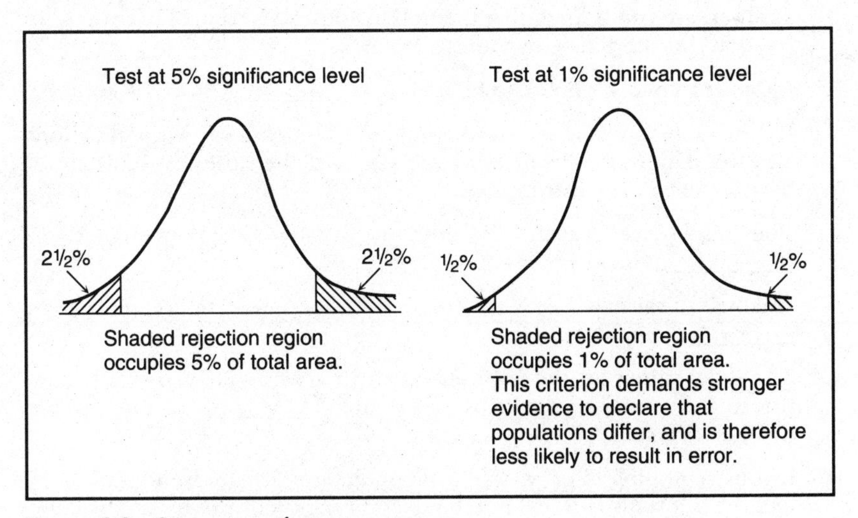

Figure 6-8. Comparison of rejection regions.

Self-Check

Answers for the following exercises can be found in Appendix A.

1. The formal statistical procedure for comparing populations is called __________ testing.
2. Which kind of error is committed when the data suggest that two populations differ but in fact they do not?
   a. type I
   b. type II
3. Which hypothesis states that two populations are alike?
   a. null
   b. alternative
4. A hypothesis test may result in an erroneous inference because it is based on data taken from __________.
5. In doing a hypothesis test the data from the samples are used to compute one number called the __________ statistic.
6. If a test statistic falls in the rejection region for a hypothesis test, the finding is called __________ significant.
7. Which test has a greater chance of resulting in a type I error, one at the 5 percent significance level or one at the 1 percent significance level?
8. If a test statistic has a low p-value does that mean there is a low or a high probability of a type I error?
9. If you set up a study hoping to show that your product is preferred to another, are you hoping that the null hypothesis is true, or are you hoping that the alternative hypothesis is true?
10. In order for a test statistic to be useful, its __________ must be known.
11. If a hypothesis test is set up to compare the means of two populations and the data comes from two large samples, then the test statistic will have a (normal, t) distribution.
12. The chi-square statistic is used to analyze data presented in __________ tables.
13. "Analysis of variance" refers to comparisons of means among more than __________ populations.
14. The set of values for the test statistic which are not likely to occur when the null hypothesis is true is called the __________ region for the hypothesis test.
15. If you are hoping to prove that two populations are different, would you like the test statistic to fall in the rejection region or not in the rejection region?

16. A matched pairs design with five pairs has more __________ than one which compares two independent samples of size five.

17. When you increase the sample size what does this do to the power of the study?
    a. increase
    b. decrease

18. In a hypothesis test comparing two populations at the 5 percent significance level, the test statistic turns out to have a p-value of .021. Is this strong enough evidence to support the claim that the two populations differ?

19. You read that a study failed to show any difference between two populations. If the study is wrong, the type of error is called
    a. type I
    b. type II.

7
Regression, Correlation, and Prediction

Key Concepts

38. Regression is a statistical technique for explaining or predicting the behavior of one variable based on information about other variables.
39. When only one predictor variable is used, the analysis is called simple regression. When more than one predictor is used, the analysis is called multiple regression.
40. The regression model is the formula which relates explanatory or predictor variables to the variable being explained or predicted.
41. The statistical analysis assigns weights (called regression coefficients) to the predictor variables used in the model. These weights help you to interpret the contributions made by their associated variables.
42. The person doing the research must have good reasons for including each potential predictor in the model because the final weights depend strongly on the set of variables presented for consideration.
43. A scattergram is a graph for showing the relationship between two variables. Looking at such a graph should be an early step in a regression analysis.
44. In simple regression, the model, called the regression line, is the straight line that overall comes closest to the dots in the scattergram.

45. The coefficients for the model are statistics based on sampling and therefore are subject to sampling error, as are any predictions based on the model. Approximate confidence intervals can be constructed using the usual rule of thumb and the standard error of estimation, which is a statistic calculated as part of a regression analysis.
46. The coefficient of determination, r^2, is a statistic calculated as part of regression analysis which is used to evaluate how well the predictor variables account for the behavior of the dependent variable. The closer r^2 is to 1, the better the model fits the data.
47. The interpretation of a regression analysis must be done very carefully, particularly when talking about causality. A statistical analysis may not explain *why* a relationship exists.

The Language of Statistics

Regression analysis

Dependent variable

Predictor variable

Linear simple regression

Multiple regression

Scattergram

Positive correlation

Negative correlation

Uncorrelated

Regression line

Slope

Intercept

Coefficient of determination, r^2

Correlation coefficient, r

Error sum of squares

Total sum of squares

Stepwise regression

Logistic regression

The Purpose of Regression

Key Concept 38

Regression is a statistical technique for explaining or predicting the behavior of one variable based on information about other variables.

Key Concept 39

When only one predictor variable is used, the analysis is called simple regression. When more than one predictor is used, the analysis is called multiple regression.

Key Concept 40

The regression model is the formula which relates explanatory or predictor variables to the variable being explained or predicted.

The purpose of regression analysis is to quantify *relationships* among variables. You may know intuitively, for example, that sales revenue is related to advertising expenses, or that the market value of a lake front vacation property depends on the length of lake frontage. Formal regression analyses of these relationships would estimate *on average* how many dollars of sales result from each dollar spent on advertising, or how many dollars each foot of frontage contributes to the market value of the property. A regression analysis produces a formula in which these relationships are explicitly displayed. This formula is called the *regression model* for the study.

Here are a few situations where a regression analysis may be helpful.

1. Determining the relationship between the amount of shelf space a product gets and the volume of sales.

2. Determining the relationship between a competitor's price and your sales.
3. Determining which variables are useful for estimating the time (or cost) to complete a project.

In the simplest case we try to illuminate the relationship between just *two* variables, say advertising expenses and sales revenue. Sometimes, however, an analysis with just two variables does not give you a realistic enough picture. For instance, to get a more complete understanding of the relationship between advertising and sales revenue, you might want to know the separate contributions of the budgets for radio advertising and newspaper advertising on sales revenue. This latter situation would lead to a more complicated regression formula. The more variables you include, the more your model may reflect reality, but as a consequence, because there will be so many interrelationships among the variables, the results of the regression analysis become more difficult to interpret. A wise goal for a regression analysis is to discover the simplest and most sensible model that adequately explains the relationships you are studying.

Dependent and Predictor Variables

Key Concept 41

The statistical analysis assigns weights (called regression coefficients) to the predictor variables used in the model. These weights help you to interpret the contributions made by their associated variables.

Basically a regression study proceeds like this. You identify the variable you are trying to explain or predict (such as gross sales). This is called the **dependent variable**. Then you draw up a list of any variables that you think might affect or influence or help determine the dependent variable. These are called the **predictor**, or explanatory or independent variables. The regression computation then assigns weights, called

regression coefficients, to each of the predictor variables according to standard mathematical criteria. You can then decide which, if any, of the predictor variables have a statistically significant relationship with the dependent variable. In this context "statistically significant" means that the correspondence is one that is stronger than would be expected by chance.

> **Key Concept 42**
>
> *The person doing the research must have good reasons for including each potential predictor in the model because the final weights depend strongly on the set of variables presented for consideration.*

One point which cannot be stressed too much but which is often forgotten is that *the weights in the final regression formula depend on the list of variables submitted for inclusion in the first place.* That is why, right from the start, you should have good reasons based on your knowledge of the business field for including or excluding variables from your analysis. For example, if the regression analysis assigns a certain weight to the radio advertising budget as a predictor of sales, you must understand that this is only in the context of the specific group of influences which were actually considered in the model. Was the television advertising budget included? What about the size of the community or the income level of the community? If one of these is a better predictor, then including it in the model would reduce the weight assigned to the radio budget.

Simple Linear Regression

Regression models can be classified by the number of predictor variables they contain and by the nature of the mathematical formula used to connect the variables. The simplest case involves just one predictor variable and a mathematical formula which corresponds to a straight line graph. This is called *simple linear regression,* and it is the most commonly applied model. When more than one predictor is involved, the analysis is called *multiple linear regression* and the formulas are straightforward extensions of those for straight lines. Most of the important points about regression can be made with the simple linear case, and so we give such an example in some detail.

The Scattergram

> **Key Concept 43**
>
> *A scattergram is a graph for showing the relationship between two variables. Looking at such a graph should be an early step in a regression analysis.*

The first step in a regression analysis is to *look at the data* by drawing a graph called a **scattergram** or scatterplot. This picture helps you to see how the values of the predictor variable correspond to the values of the dependent variable. It is very risky to interpret a regression analysis without this picture.

Example 7-1: Mail Order Sales. The manager of the shipping department at a mail order company has reason to believe that the number of orders that arrive in the first mail delivery on Monday morning will be a good predictor of the total number of orders for the week. Having this information early in the week facilitates efficient assignment of workers. To investigate this relationship, data is collected for 30 weeks and recorded as shown in Table 7-1. This data is displayed in the scattergram of Figure 7-1.

In a scattergram the usual procedure is that the vertical axis represents the dependent variable, and the horizontal axis represents the predictor variable. Each single dot represents a *pair* of numbers. Since there are 30 pairs of values in the data of Table 7-1, there are 30 dots in the graph. For instance, the pair of values for week 1 is (94, 1181) and so there is a single dot (which is circled) located at the position which is over the 94 and to the right of 1181 (or as close to that position as the computer can show it).

The first impression the graph gives is that the whole array of dots slants upward to the right. This happens because larger orders on Monday morning *do* roughly correspond with larger total orders for the week. Such a pattern is called **positive correlation**. Two variables are positively

Table 7-1. Mail Orders Received at a Company Over a Period of 30 Weeks

Week	1	2	3	4	5	6	7	8	9	10	11	12	13	14	15
Monday A.M. orders	94	86	114	123	97	80	82	77	96	106	121	78	61	91	117
Total for the week	1181	1320	1685	1712	1527	1269	1141	1103	1535	1590	1608	1205	1033	1415	1735
Week	16	17	18	19	20	21	22	23	24	25	26	27	28	29	30
Monday A.M. orders	119	128	91	116	86	108	85	86	83	73	123	131	106	93	70
Total for the week	1860	1825	1512	1624	1242	1630	1246	1369	1277	995	1658	2059	1670	1274	1089

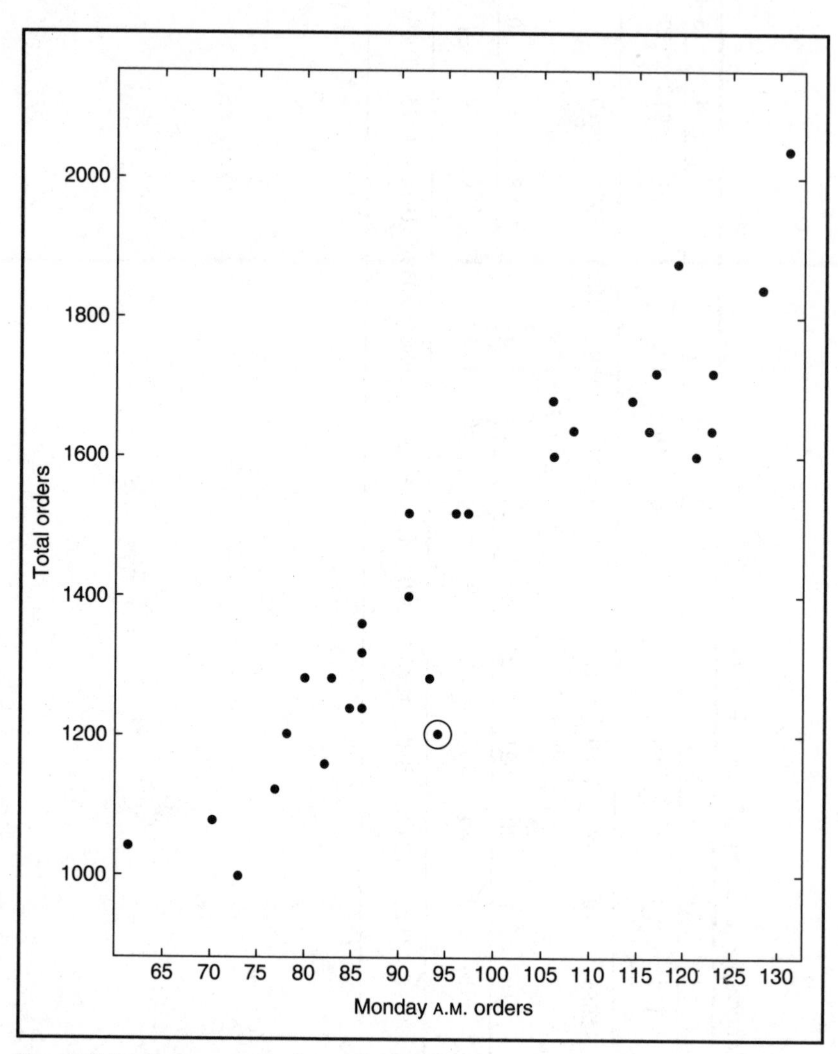

Figure 7-1. Scattergram for data in Table 7-1.

correlated when increases in one tend to be associated with increases in the other. Conversely, two variables are **negatively correlated** when increases in one tend to be associated with decreases in the other, in which case the array of dots slants *downward* to the right. In a totally random relationship the two variables are called uncorrelated. See Figure 7-2 for a summary sketch.

The Regression Line

Key Concept 44

In simple regression, the model, called the regression line, is the straight line that overall comes closest to the dots in the scattergram.

When you look at a scattergram as if it were a kind of Rorschach test, you may think that the dots are trying to make a simple pattern, perhaps a line or curve. Statisticians assume that if there were not random inexplicable forces in the world, then the graph would, in fact, form some ideal shape which reveals the relationship between the predictor and dependent variables. The purpose of the regression analysis is to describe this ideal shape—the perfect figure the dots are "trying" to make. Each ideal shape has its own formula, and the formula for a straight line is the simplest. Consequently, simple linear regression starts by *assuming* that the dots are trying to be in a straight line and then uses the mathematics of regression analysis to come up with the exact formula for this particular ideal line. The regression calculations determine the particular line which overall comes closest to all the dots in the scattergram. It is called the **regression line** or the **line of best fit**.

The mathematical formula for a line is characterized by two coefficients, the **slope** and the **intercept**. The slope describes the steepness of the line, and the intercept gives the value of the dependent variable

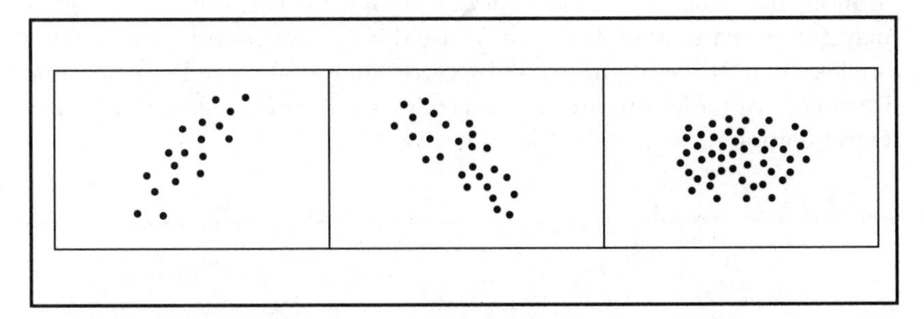

Figure 7-2. Positive, negative, and zero correlation.

when the predictor is equal to zero. If we perform a regression analysis on the data in Table 7-1, in order to fit a straight line to the data we will find that the slope is about 13.4 and the intercept is about 140. In Figure 7-3 we have superimposed the graph of this line on the scattergram.

The Formula for the Regression Line

The formula for the regression line has the form

Expected value of variable to be predicted

= intercept + (slope × value of predictor variable)

In this particular example the formula would be

Expected total orders for week

= 140 + (13.4 × number of orders on Monday morning)

The slope value 13.4 means that for each additional order on Monday morning, on average there are 13.4 additional orders for the week. In general, the slope of the regression line tells you how many units the dependent variable is expected to change when the independent variable changes by *one* unit. The value of the intercept gives the average value of the dependent variable when the predictor variable equals *zero*. This must be interpreted with care as it is sometimes not a sensible concept, particularly if zero is far below the smallest observed value for the predictor variable. In our present example, the lowest value for Monday A.M. orders was 61. The intercept implies that even with zero orders on Monday morning we should expect 140 sales for the week; this may or may not be reasonable. It is risky to use this line for Monday sales which are less than 61. Going beyond observed values of the predictor variable is called extrapolation, and there is always increased risk of error with extrapolation.

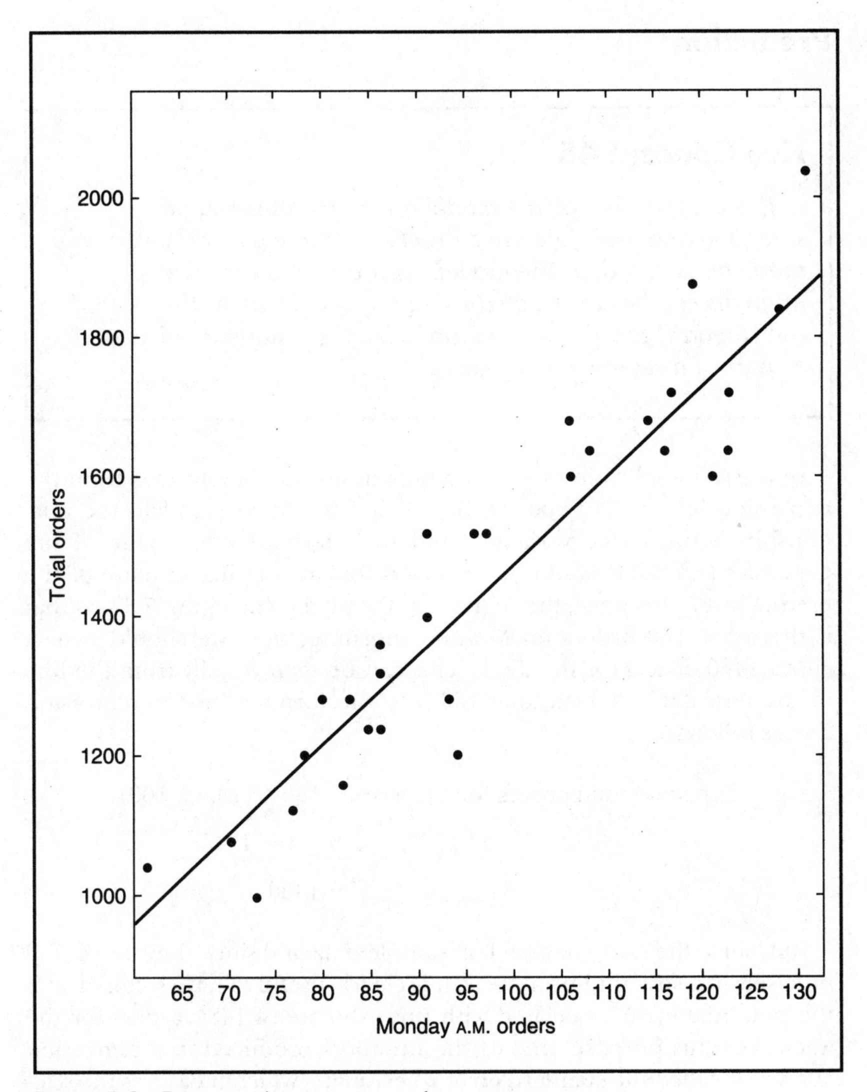

Figure 7-3. Regression line superimposed on scattergram.

Prediction

Key Concept 45

The coefficients for the model are statistics based on sampling and therefore are subject to sampling error, as are any predictions based on the model. Approximate confidence intervals can be constructed using the usual rule of thumb and the standard error of estimation, which is a statistic calculated as part of a regression analysis.

In a given application when the dots do not deviate too far from the regression line, we may be satisfied to use the line to explain the relationship between the variables, and to determine what value of the dependent variable would be expected for any particular value of the predictor. For instance, the regression formula from Figure 7-3 says that if there are 100 orders on Monday morning, then we should expect about 1480 orders for the week. This can be seen visually from the line or numerically by substituting 100 into the formula for the regression line as follows.

$$\begin{aligned}\text{Expected total orders for the week} &= 140 + (13.4 \times 100)\\ &= 140 + 1340\\ &= 1480\end{aligned}$$

But, since the dots are based on sample data and since they do not fall precisely in a straight line anyway, it is clearly useful to have some idea of the potential error associated with this estimate of 1480 orders for the week. For this purpose, one of the numbers produced in a regression analysis is called the **standard error of estimate**, which measures the scatter of the dots around the regression line. (It would be zero if the dots fell perfectly in line.) As we described in the chapter on estimation, a reasonable rule of thumb for 95 percent confidence is to use twice the standard error of estimate as a margin of error for predictions of the dependent variable. For the regression line of Figure 7-3, the standard error of estimate is 97.5, and so the margin of error is two times 97.5 or 195. An approximate 95 percent prediction interval for the total weekly orders based on 100 Monday morning orders then would be 1480 ± 195.

To get more precise calculations of the prediction interval much more complex formulas are required. Most commercial statistical software will do the correct calculations for you. Computer calculations for the prediction above yield a margin of error equal to 203. You should also note that it is a property of the regression analysis that estimates based on values of the predictor near the mean have smaller error than those based on smaller or larger values. This supplements the earlier warning that extrapolation beyond the range of observed values of the predictor is risky.

How Well Can You Expect the Line to Work?

Surprisingly, the mathematics which determines the regression line is blind to the actual pattern of the dots. You can start with any pattern of dots, and the calculations will find the one particular line that overall comes closest to all the dots. However, if in fact, the pattern of the dots is not close to a straight line, the regression line will be worthless for prediction and estimation. This is one reason for *looking* at the scattergram before you do a regression analysis.

Key Concept 46

The coefficient of determination, r^2, is a statistic calculated as part of regression analysis which is used to evaluate how well the predictor variables account for the behavior of the dependent variable. The closer r^2 is to 1, the better the model fits the data.

To supplement your visual impression, you can use a numerical measure of how close the dots are to being in a straight line. One statistic calculated during a regression analysis is a number called the **coefficient of determination** which is usually symbolized as r^2 and which is a measure of how well the model fits the data. The numerical value of r^2 is always some number from 0 to 1, where 1 indicates a perfect fit and 0 indicates a poor fit.

In any regression analysis—no matter how complex the model is—you should always find out the value of r^2; it gives you a rough idea of how well your set of predictor variables accounts for the dependent variable.

In the particular case of simple linear regression, where there is only one predictor and one dependent variable, it is common also to use the square root of r^2 as a measure of how close the dots come to being in a straight line. The statistic is just symbolized as r, and is called the **correlation coefficient** between the two variables. In agreement with our earlier discussion of correlation, r is given a positive sign if the line slants upward to the right, and a negative sign if the line slants downward to the right. If the best fitting line is horizontal, both the slope and the correlation coefficient are 0.

For the regression line in Figure 7-3 calculations give the value of r^2 as 0.877, and the value of r as 0.936, both of which, intuitively, are close to 1, indicating that the regression line does a good job of describing the pattern of dots. The simple linear model fits the data pretty well, and Monday A.M. sales do serve as a reasonable predictor for total weekly sales. Note: Some people prefer to report the coefficient of determination as a percentage. For this data they would say that r^2 is 87.7 percent.

More Detail on the Value of r^2

People who routinely use regression analyses as part of their research develop an intuition for how well models usually fit the particular kind of data they work with. They see similar values for r^2 in many studies. As a consequence, they develop a gut feeling for when a model fits the data well. In a controlled manufacturing laboratory where everything that contributes to a particular outcome is known and accurately quantified, values of r^2 above 0.90 may be frequent. By contrast, analyses based on "real world data," which are subject to many unknown influences, as in a survey designed to predict human or animal behavior, may typically yield values of r^2 around 0.40. Experienced workers in a given field know what to expect and would very quickly spot a report that seemed out of line. For example, a report which had a much higher value of r^2 than usual might mean a major breakthrough in the field; it could mean that researchers have new insight into what determines some particular outcome. Or, as some disgraced scientists have learned, it might be a sign of fraud, of data that were tampered with or made up to prove a point. More than one scientist has been trapped by made-up data because correlation coefficients were suspiciously high.

In the next several paragraphs we give a more precise interpretation of the numerical value of r^2. You may find the discussion harder to follow because it does not have a simple intuitive explanation. If you choose to not bother with the details, just remember that a value of r^2 close to 1 means a good fit and a smaller error of estimation. The basic idea is that the value of r^2 represents the degree to which using the given set of predictors serves as an improvement over using *no* predictor variables in the model.

Suppose we decide to use Monday A.M. sales to predict total weekly sales, and suppose that there really is no connection, that the Monday sales are in fact useless as a predictor of total weekly sales. Another way of saying this is that there is no correlation between the two variables. When this is true, the regression line is horizontal, with intercept equal to the overall average for total weekly sales. The formula would look like this.

$$\text{Weekly total sales} = \text{mean weekly sales} + (0 \times \text{Monday A.M. sales})$$

This model says that no matter what the sales are on Monday morning, just predict that the total for the week will be the mean of all weekly sales recorded in the data set. For the data set of Table 7-1, the mean weekly sales is 1446.3 and this line is shown superimposed on the data set in Figure 7-4.

This line will miss each dot by a certain distance. A measure of the adequacy of this model can be computed based on all these distances. This measure is often called the **total sum of squares** for the data. It is found by simply taking the sum of the squares of the distances by which the line misses each point. In our example, the sum of the 30 squared distances equals 2,162,934.3. This number serves as a bench mark against which to compare models which do include the correlation between the predictor and the dependent variable. When this correlation is included in the model, the value of the slope is not zero, because the regression line moves and tilts to get closer overall to the dots. Even so, this better regression line will also miss each dot by a certain distance, and we can compute a measure of total error for this line as well. This is often called the **error sum of squares**, and it will be a smaller value than the total sum of squares. The calculations of the regression analysis always determine that particular line which yields the smallest possible error sum of squares. For the line in Figure 7-3 the value is 266,310.5.

The ratio $\dfrac{\text{error sum of squares}}{\text{total sum of squares}}$ therefore can serve as a measure of how closely the regression model fits the dots. The better the fit, the smaller is the numerator, and the closer the ratio comes to zero. For our data set the ratio is .123 or 12.3 percent. This represents the degree to which the regression line *failed* to account for the behavior of the dependent variable; it is the error which remains after the regression line is drawn. To describe the *success* of the regression line we can subtract this number from 1. The result of the subtraction is what we have called r^2.

$$r^2 = 1 - \frac{\text{error sum of squares}}{\text{total sum of squares}}$$

When the regression line fits the data well, r^2 is near 1.

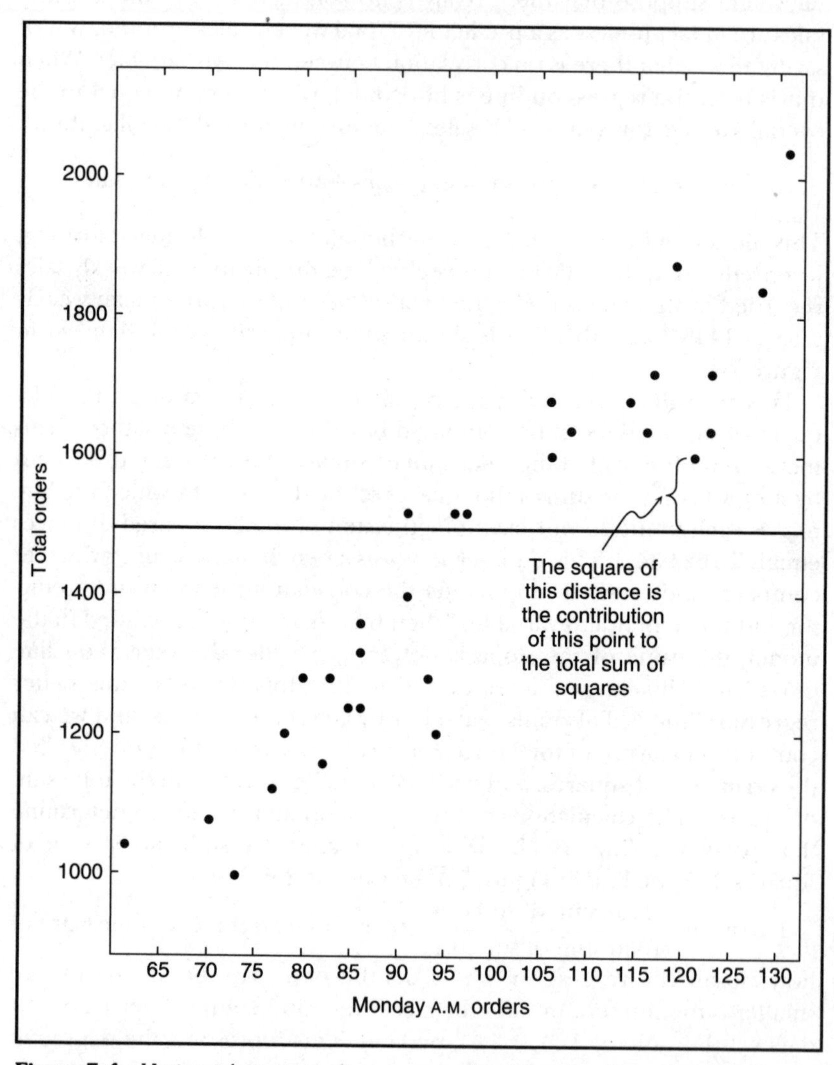

Figure 7-4. Horizontal regression line corresponding to no correlation between Monday A.M. sales and total weekly sales. Intercept equals the mean of weekly sales, 1446.3.

For the regression line in Figure 7-3, r^2 is 0.877. Using the predictor variable then reduces the overall error to 12.3 percent of the total sum of squares, a marked improvement. In the jargon of statistics, we say that the model which includes the predictor "accounts for" or "explains" 87.7 percent of total sum of squares. In fuzzier language people say that the predictor accounts for 87.7 percent of the variability of the dependent variable. The other 12.3 percent is called "unexplained," reflecting unaccounted for influences on the dependent variable.

Interpreting Correlation

Key Concept 47

The interpretation of a regression analysis must be done very carefully, particularly when talking about causality. A statistical analysis may not explain why *a relationship exists.*

It is important not to get carried away with the promise of regression to make predictions. Remember that regression and correlation are mathematical concepts which describe patterns of numbers. They don't always illuminate what is really going on in the real world. It is possible, for instance, that you may find out that one variable is positively correlated with another and still have no idea why. And when you don't know why, you can't sensibly interpret the relationship, and you won't know how to act upon this information. The well-known catch phrase is "correlation is not causation." Quantifying a statistical relationship should not be a substitute for understanding. To illustrate the point consider the data in Table 7-2 which are taken from the 1990 issue of the *Statistical Abstract of the United States.*

A scattergram is shown for this data in Figure 7-5. For this data there is a rather high positive correlation. The value of r is 0.897 and r^2 is 0.804. But it is not clear why this correlation exists, nor what to make of it. In a statistical sense this says that you could use the number of births to predict sales of imported automobiles. But why? Is it just a fluke? Are both sets of figures responding to some larger social phenomenon such as the general economic climate? The regression analysis by itself will not answer this substantial question.

Table 7-2. Number of Births in the United States (in thousands) and Value of Imported New Automobiles (in millions of dollars) from 1980 to 1988

Year	Births	Value of imported new automobiles
1980	3612	16766
1981	3629	17540
1982	3681	19757
1983	3639	22934
1984	3669	29208
1985	3761	36475
1986	3757	45302
1987	3809	47858
1988	3918	47005

In contrast to this example, there may be a situation where you do understand the connection between two variables but you do not have the data at hand to quantify the relationship. This is an opportunity to set up a research project to ascertain the data for a regression analysis. For instance, unless there is a radical change in vision care, it is likely that the demand for bifocals will follow the size of the population that is in the over-40 age group. Since that covers the "baby-boomers" one can safely predict in a general way that sales of bifocals will be increasing during the 1990s. You would expect the data to show a positive correlation between the number of births and the number of bifocals sold, say, 45 years later. Then you could use regression analysis to make more specific predictions.

Multiple Regression

As noted earlier there are circumstances where simple regression is inadequate, where multiple regression is more sensible. For example, in analyzing the sales of a given product or service you may be able to draw up a list of *several* factors which would reasonably be thought to influence total sales in a particular region. These might include the size and wealth of the region, the price of the product or service, and some measure of the competition in the region to name a few. If you had data for all these variables you could include them all as potential predictor variables for your model. Then you could use appropriate statistical tests to eliminate those which are redundant and do not therefore make a statistically sig-

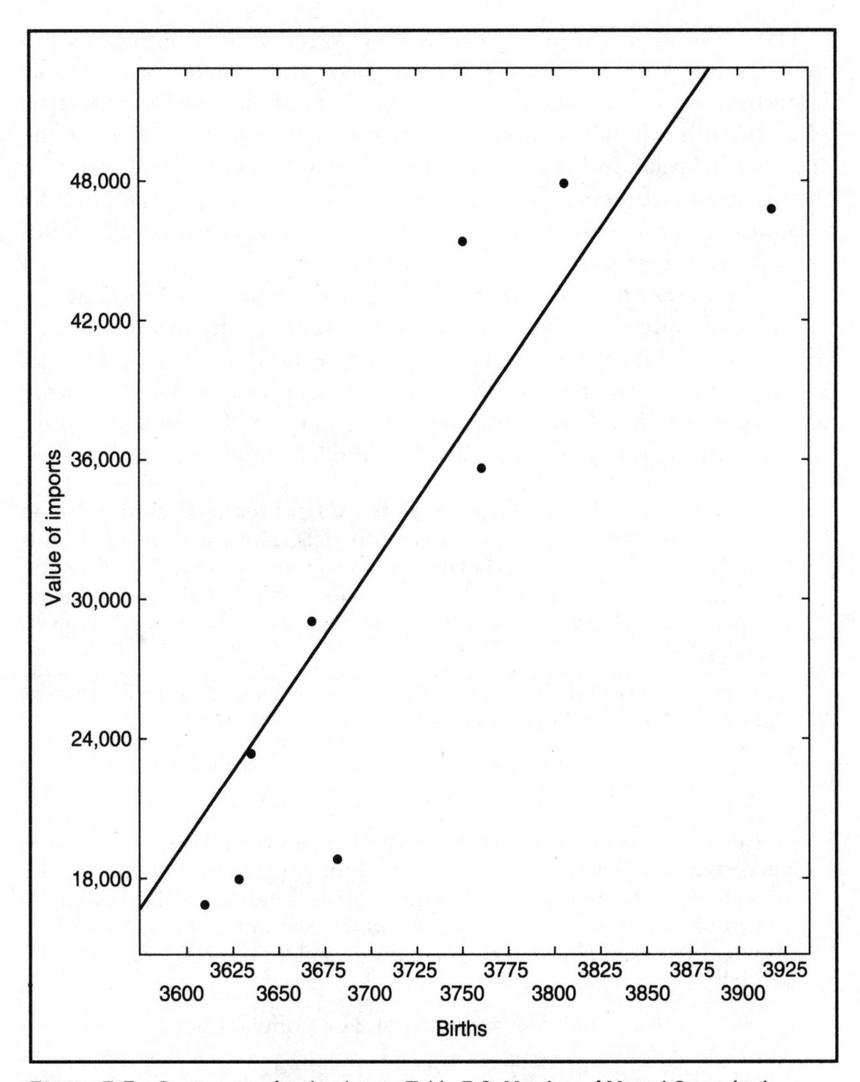

Figure 7-5. Scattergram for the data in Table 7-2. Number of United States births (thousands) and value of imports of new automobiles (millions of dollars). 1980 to 1988.

nificant contribution to estimating sales beyond what the other predictors can achieve. *The goal is to come up with the simplest model that reasonably explains what determines sales.* In this context, to "explain" how a predictor influences sales means to establish a numerical relationship between the value of that predictor and total sales.

The whole business of regression rapidly gets more complex as you increase the number of predictor variables, and it takes considerable expertise to do a good job of it. As we have mentioned before, this type of analysis should not be done by someone who doesn't appreciate the "real world" relationships among the variables; this is an ideal arena for cooperation between a professional statistician who understands the complexity of the mathematical model and a manager who understands what the numbers stand for in the marketplace.

This book is not intended to go deeply enough into the matter to cover all the questions that must be answered along the way to building a reasonable regression model involving several predictors. Instead, showing the results of one such analysis, which is assumed to be reasonably constructed, will be adequate. The interpretation of the results, rather than the process of building the model, will be emphasized.

Example 7-2: Using Two Predictor Variables to Estimate Sales Volume. A firm operates in ten sales districts. The sales manager wants to quantify the relationship between sales volume in the various sales districts and two predictor variables, population in the district and the average per capita income in the district. Let us suppose the data are as given in Table 7-3.

The simplest multiple regression model is a straightforward extension of the simple linear regression model and would take the form

$$\text{Expected sales} = \text{intercept} + (\text{weight} \times \text{population}) + (\text{weight} \times \text{per capita income})$$

Recall that the general name for these weights and the intercept is regression coefficient. As in the case of simple regression, submitting the data above to a multiple regression analysis determines the numerical values of the regression coefficients which minimize the error sum of squares. For the data in Table 7-3 the calculations yield the following model.

$$\text{Expected sales} = -40.2 + (1.34 \times \text{population}) + (2.38 \times \text{per capita income})$$

As part of the output we would see that r^2, the coefficient of determination, is equal to 0.944 or 94.4 percent, and that the standard error of estimate is 1.90.

This model can be then be used to predict sales when the population and per capita income for a sales district are both given. For example, in a region where population is 30 thousand people and per capita income is 15 thousand dollars,

$$\text{Expected sales} = -40.2 + (1.34 \times 30) + (2.38 \times 15) = 35.7 \text{ units}$$

Table 7-3. Sales of a Product for Various Sales Districts

District	Sales last year (product units)	Population (thousands)	Per capita income (thousand dollars)
1	33.3	32.4	12.5
2	35.5	29.1	16.2
3	27.6	26.3	14.5
4	30.4	31.2	13.1
5	31.9	29.2	13.1
6	53.1	40.7	15.8
7	35.6	29.8	14.9
8	29.0	23.0	15.2
9	35.1	28.2	16.2
10	34.5	26.9	15.7

As in the one predictor case, we can use the value of r^2 to indicate how closely the models fits the data. In this example the value 0.944, because it is quite close to 1, does indicate that the model fits the data well. Further, we can use the standard error of estimate to get a rough margin of error for any particular estimate. Using twice this standard error as a rule of thumb, the approximate margin of error is about twice 1.90, or 3.8 units of sales. (At the end of the chapter, in the section called "Some technical issues," there are some remarks about the accuracy of this rule of thumb.) To interpret the weights in the model, recall that population was recorded in thousands of people, and that per capita income was given in thousands of dollars.

Weight on population = 1.34. For every one unit increase in population (that is, for every increase of one thousand people) we can expect on average an increase in sales of 1.34 units.

Weight on income = 2.38. In addition to the influence of population, for every one unit increase in per capita income for the sales district (that is for each one thousand dollar increase) we can expect on average an increase in sales of 2.38 units.

Intercept = –40.2. This is just the mathematical value that the

Table 7-4. Correlation Table for the Data in Table 7-3

Correlations:	Sales last year	Population	Per capita income
Sales last year	1.0000	.8552	.4002
Population	.8552	1.0000	–.0696
Per capita income	.4002	–.0696	1.0000

model produces when the population is zero and the per capita income is zero. It has no meaning as a sales figure in this context.

Often the output for a multiple regression analysis includes a table of correlation coefficients for the variables in the model. The correlation table for this example is shown in Table 7-4, where each entry is the correlation coefficient between the two variables named at the side and top. The values will always be 1 along the diagonal, since a variable is always perfectly correlated with itself.

These values indicate, as expected, a strong correlation (.8552) between the size of the population in a sales district and the total annual sales there, and a weaker correlation (.4002) between per capita income and sales. The correlation between the two predictors is close to zero (–.0696).

It is fortunate in multiple regression when the predictors are not correlated with one another; this allows them each to make an independent contribution to the explanation of the behavior of the dependent variable. When there is high correlation between the predictors, it is harder to interpret the independent contributions of the various predictors. This situation is called **multicolinearity**, and should be avoided by careful choice of the list of potential predictors. Looking at the table of correlations in order to spot likely candidates for predictors is a good early step in multiple regression, just as looking at the scattergram is a good early step for simple regression.

Multiple Regression in Stages—Stepwise Regression

A more detailed output from a multiple regression analysis may indicate the method by which the weights were computed. A commonly applied method is called **stepwise**, by which first one then another predictor is entered into the model. This allows you to decide at each stage if adding another predictor variable makes a significant improvement to the reliability of the model. Once more, because this can be a complex matter, you may want some expert guidance. For the sales data above, the following results are what you would see at each stage.

Stage 1. Because population is the predictor with the strongest correlation to sales, it will be the first predictor included in the model, and the calculations lead to:

Model at stage 1: Expected sales = $-3.75 + (1.29 \times \text{population})$

For this model the value of r^2 is .731, and the standard error of estimate is 3.89.

Stage 2. By going on to stage 2 where per capita income is also included in the model, we get the model shown earlier:

$$\text{Model at stage 2: Expected sales} = -40.2 + (1.34 \times \text{population}) + (2.38 \times \text{per capita income})$$

When we include the second predictor, r^2 increases to .944 and the standard error drops to 1.90. The model fits the data better and the margin of error for estimates is smaller. Most experts would agree that using the stage 2 model gives a statistically significant improvement. (This can be shown by a formal hypothesis test.) But, more importantly, *it makes business sense,* that per capita income influences sales independently of population.

Some Technical Issues

In order not to interrupt the discussion at too many points, we have glossed over some important points.

1. Confidence intervals for the regression coefficients. The weights produced by the analysis are based on sample data. They are therefore statistics which are estimates of the "true" weights, in some perfect ideal population model. If a weight is very close to zero, this may indicate that the corresponding predictor variable in the true population model deserves *no weight.* In short, it doesn't really belong in the model. A typical regression analysis will give a t-statistic and a corresponding p-value for each regression coefficient in the model to test the hypothesis that the true value of the coefficient in the population is zero. Large values of t, or correspondingly, small values of p indicate that the true value of the regression coefficient is not zero, and that the predictor should be kept in the model. For stepwise regression, a computer will have some default value which it uses to decide whether a predictor should be kept in the model.
2. The nature of the dependent variable. The regression models we have discussed are appropriate for dependent variables which are approximately normally distributed. We used this implicitly in the rule of thumb for the confidence interval estimates. In particular, these models are not appropriate for "yes-no" type dependent variables, which is the type you would have, say, if your purpose were to predict whether or

not a customer will buy a product, or an employee will quit a job. These models should be analyzed differently; often the correct approach is one called **logistic regression**, which we won't discuss in this book, but only note as appropriate for yes-no dependent variables.

3. The accuracy of estimates or predictions. We have given an approximate rule of thumb, which says that the margin of error for estimating the value of the dependent variable is about twice the standard error of estimate. The calculations for more exact intervals are complicated and are best done by computer. Consider that the rule of thumb is a ballpark estimate of the correct order of magnitude. The next paragraph discusses the exact calculations, which would normally be done by computer.

   In the context of regression, statisticians distinguish between a **prediction interval** and an **estimation interval**. These answer two subtly different questions and they have two different standard errors. Our rule of thumb produces an answer somewhere between them. The estimation interval answers the question: What is the *mean value* of the dependent variable for *all* the members of the population at any given set of values of the predictors. For instance, what is the mean value for sales in all districts with population 30,000 and per capita income $15,000. The prediction interval answers the question: What is the *best guess* for sales in *one* district with a given set of predictor values. For instance, what is the best guess for sales in a district with population 30,000 and per capita income of $15,000? For both questions the point estimate is the same; in our example it would be 35.7 units. The margin of error is larger for the prediction interval, however, because it is an estimate for a single case rather than a mean of many cases. Our rule of thumb gave a margin of error equal to 3.80 units. Exact calculations give the margin of error for the estimation interval equal to 1.46 and for the prediction interval equal to 4.73.

Self-Check

Answers to the following exercises can be found in Appendix A.

1. In simple regression, the number of predictor variables is __________.
2. In a regression analysis the variable you are trying to predict is called the (dependent, independent) variable.
3. A formula that shows the relationship between predictor and dependent variables is called a regression __________.

4. The general name for a weight assigned to a predictor variable in a regression model is "regression ________."

5. A graph made up of dots which shows the relationship between two variables is called a ________.

6. Draw a scattergram for this set of data.

Predictor	4	5	6	7	8
Dependent	10	9	7	6	2

7. If two variables are positively correlated, then increases in one are associated with ________ in the other.

8. If the slope of a regression line is equal to 8, then on average we expect a change of 1 unit in the ________ variable to be associated with an 8 unit change in the ________ variable.

9. Would you expect negative or positive correlation in a community between the unemployment rate and sales of new cars?

10. Why is the regression line also called the line of best fit?

11. Does a regression model fit the data better when the value of r^2 is 0.86 or when it is 0.14?

12. If the value of r^2 is 0.72, then the ________ variable accounts for 72 percent of ________ of the ________ variable.

13. A regression analysis results in this model:

$$\text{Expected sales units} = 400 + (20 \times \text{radio ad budget}) + (15 \times \text{newspaper ad budget})$$

If the standard error of estimate is 10 units, what is an approximate 95 percent prediction interval for sales units when the radio ad budget is 50 and the newspaper ad budget is 40? Assume dollar figures are in thousands of dollars.

14. Referring to the sketch of the regression line, estimate Y when $X = 10$.

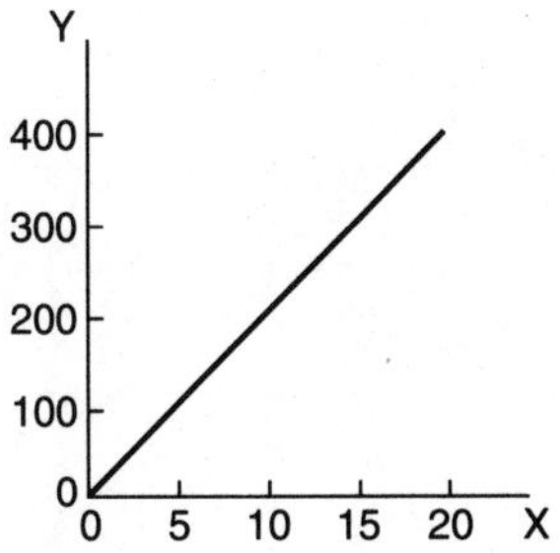

Y
400
300
200
100
0
0 5 10 15 20 X

15. If the correlation between two variables is given as –0.88, then increases in one variable tend to be associated with ________ in the other.

16. For constructing a multiple regression model, it is preferable if the predictor variables (are, are not) correlated with one another.

17. "Stepwise" regression is a procedure for building a model in (simple, multiple) regression.

18. In stepwise regression, the predictor variable that is the first one to be included in the model is the one with the (highest, lowest) correlation with the dependent variable.

19. An advertising campaign is designed to try to get homeowners who heat with oil to switch to gas heat. Several predictors will be used, and the dependent variable is a person's response to the question "Did you switch?" Because this is a yes-no question, the best approach is not linear multiple regression, but is __________.

20. Do the data graphed in this scattergram indicate that a linear regression model would be useful for predicting the dependent variable based on this predictor variable?

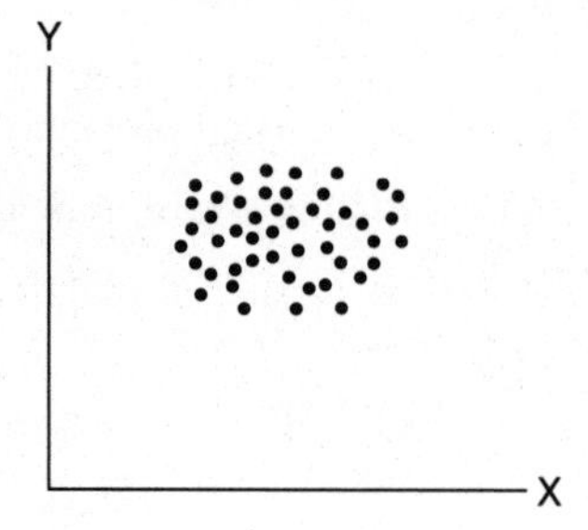

8

Times Series: Forecasting and Interpreting Trends

Key Concepts

48. A time series is a set of values for a variable recorded over time.
49. A model for a time series which takes account of correlation between values within the series is called autoregressive (AR).
50. A model which shows how each observation in a time series depends on the ones previous to it is called a moving average model (MA).
51. Sophisticated business time series analysis is often done using ARIMA models (also called Box-Jenkins), which are both autoregressive and moving average.
52. The four components of a time series are secular trend and the seasonal, cyclical, and irregular components.
53. Decomposition of a time series is a procedure for identifying and describing each of the four components of a time series.
54. Seasonal adjustment of a time series means removing the seasonal component.
55. Smoothing a time series is a technique for revealing underlying patterns. Two elementary smoothing techniques are the moving average and exponential smoothing.

56. To get the seasonal forecast multiply the underlying trend value by the seasonal index.
57. Two popular statistics for evaluating how well a smoothed series mimics a given time series are the mean squared error (MSE) and the mean absolute percentage error (MAPE).

The Language of Statistics

ARIMA model
Autocorrelation
Autoregressive model
Box-Jenkins model
Centered moving average
Components of a time series
Cyclical component
Decomposition of a time series
Exponential smoothing
Forecast
Irregular component
MAPE
Moving average
MSE
Seasonal component
Seasonal index
Seasonal adjustment
Secular trend
Simple exponential smoothing
Smoothing constant
Smoothing a time series
Time series

Introduction

Key Concept 48

A time series is a set of values for a variable recorded over time.

An essential task of management is to follow the progress of business performance over time. One statistical approach to this task is the recording and analysis of **time series**, which is just the technical name for a set of values for a variable recorded over time. The distinguishing feature of a time series is that each observation is clearly associated with a specific instant in time. As a matter of common practice, the intervals at which the observations are made are usually regular, often monthly, quarterly, or yearly. Some familiar examples are monthly sales, quarterly interest rates, annual profits, and daily absenteeism. It is natural to describe a time series graphically, where the flow of time is represented by the horizontal axis as shown in Figure 8-1. The data for this graph are given in Table 8-1.

Statistics offers two complementary contributions to the analysis of time series. First is the *description of the pattern* already recorded. For example, it may become very clear through analysis of a time series that a regular pattern of variation is present. Examples would be weekly rises in absenteeism every Monday, or annual jumps in toy sales before Christmas. Understanding these regular jumps or dips in the long term pattern can shed light on the overall behavior of the series, which can help reduce surprises, and provide information for planning.

A second contribution of statistical analysis is the *projection of the series* into the future, which would usually be called business **forecasting**. The main point here should be obvious: there is no way to forecast infallibly, no matter how fancy or how detailed the technical aspects of the projection are. There are assumptions in any forecasting scheme, the main one being that the forces that generated the data in the past will continue to work in the future, and if reality decides to depart from these assumptions, then the forecasting will go wrong. As in other applications of statistics (maybe more so), the use of time series analysis for forecasting must include large doses of common sense and awareness of the forces that affect the variable of interest. By providing quantitative support for guesses about the future, time series analysis makes a valuable contribution, particularly in keeping more intuitive projections from going too far afield.

Table 8-1. Annual Production of Tobacco, United States: 1980 to 1989

Year	Production (millions of pounds)	Year	Production (millions of pounds)
1980	1786	1985	1511
1981	2048	1986	1164
1982	1994	1987	1189
1983	1429	1988	1370
1984	1728	1989	1414

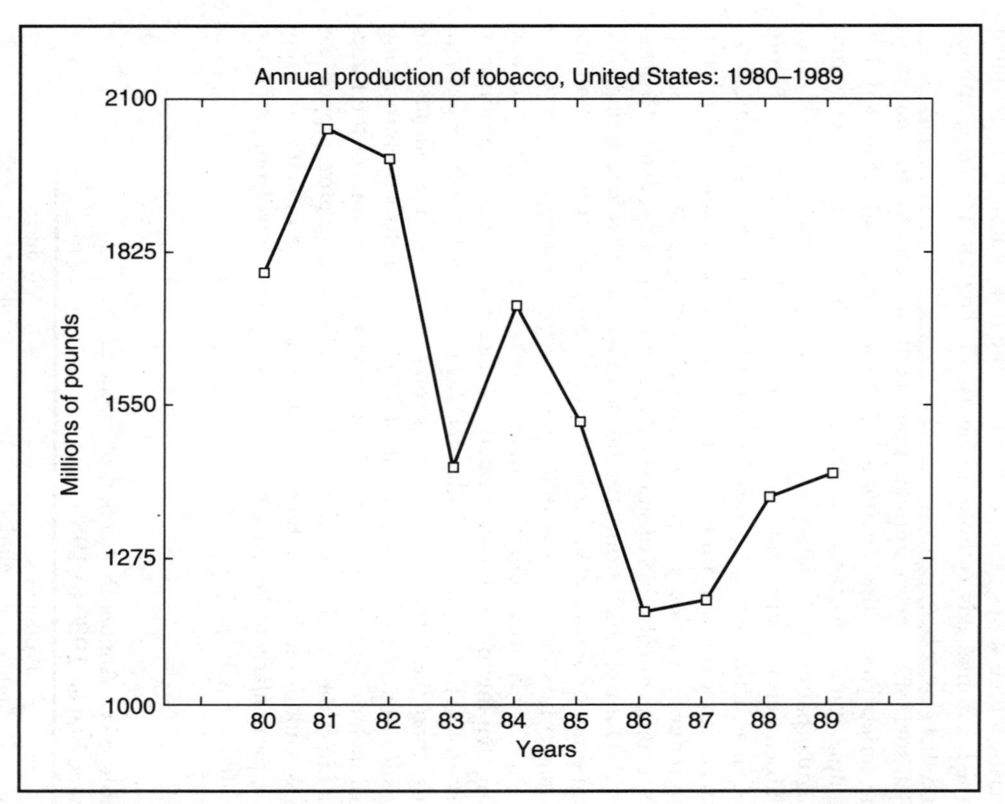

Figure 8-1. Typical time series plot.

Overview of Methods for Analyzing Time Series

Key Concept 49

A model for a time series which takes account of correlation between values within the series is called autoregressive (AR).

Key Concept 50

A model which shows how each observation in a time series depends on the ones previous to it is called a moving average model (MA).

Key Concept 51

Sophisticated business time series analysis is often done using ARIMA models (also called Box-Jenkins), which are both autoregressive and moving average.

Methods for analysis of time series range in sophistication and difficulty from simply drawing a graph and looking at it through fairly simple numerical ways of looking for patterns, to determination of quite complex models which attempt to account in one shot for all the behavior of the series. This book will concentrate on the simpler procedures, mainly the **decomposition of time series** and simple **smoothing techniques** based on **moving averages** in order to get a feeling for the subject. I'll avoid the details of the more sophisticated approaches because it takes considerable study to really make sense of them.

Generally speaking, a professional analysis today will be based on regression analysis. One approach will include not only data from the series to be analyzed, but will also include other predictor series. For instance, one might want to analyze the pattern of new housing starts in some region, and it might be helpful to include the pattern of mortgage interest rates as a predictor. Clearly, other predictor variables could also be included. Such analyses are basically concerned with creating a useful model of predictors. Economists in particular do a lot of work in this field trying to describe the links among various economic series.

Another approach is concerned only with the data from *one* series of observations. To forecast housing starts this way the analyst just looks at the pattern of housing starts over a period of time and tries to describe its ups and downs as sensibly as possible. Here the regression idea is more directed to analyzing how any particular observation depends on previous observations within the same series. The analysis will specifically take note of the fact that there is likely to be correlation between certain observations within the series. For instance, high unemployment one month is much more likely to be followed the next month by high unemployment than by low unemployment. The attempt to describe these internal relationships is called **autocorrelation analysis**, and regression models which take autocorrelation into account are called **autoregressive models**.

A full blown technique which is currently popular for analyzing time series uses complex models called "autoregressive, integrated, moving average," or **ARIMA** for short. Autoregressive refers to the need to account for the potentially high degree of correlation between observations *within* the same time series, while "integration" and "moving average" refer to aspects of the models which account for long-term trend and different degrees of influence of past values of the series on present and future values. The pioneer work was done by George Box and Gwilym Jenkins, and so this approach is also called the **Box-Jenkins** approach to time series analysis. The theory behind this analysis is not simple, and the calculations are considerable. Analysis of time series by ARIMA models should be done by those who have had enough practice to really know what they are doing.

Time series analysis and business forecasting is one area of statistics where you can easily be overcome by technicalities and lost in a sea of mumbo-jumbo. For a business manager who makes decisions based on such analyses the most important question must always be "Does this make sense?" Above all, no matter how fancy they are, don't expect forecasts to be accurate very far into the future.

The Components of a Time Series

Key Concept 52

The four components of a time series are secular trend and the seasonal, cyclical, and irregular components.

It is common in business applications to think of a time series pattern as potentially having the following four distinct components.

1. A slowly changing long-term trend, often called a *secular trend.* This is usually postulated to be an increasing or decreasing trend, often fairly well described by a straight line.
2. A *seasonal* component with a regular known period (most often one year). This component describes regular variations within a year, like predictable slack and busy sales "seasons." When the data are recorded monthly, then "seasonal" basically means "monthly." It is assumed, for instance, that there is something about all Decembers that is the same from year to year. In the same way, when data are recorded quarterly, then "seasonal" means "quarterly."
3. A more or less regular oscillation around the trend, called the *cyclical* component. This is usually intended to describe general cycles in the economy at large, such as recessions, which occur slowly over long periods of time. There is no expectation that the cyclical component should correspond to the calendar year. Identification of long-term cycles is more likely to be of interest to the government economist than to the business manager seeking shorter term projections.
4. An *irregular* component, consisting of short-term fluctuations of unexplained nature.

These four components provide a convenient way to describe time series so that they make sense, but as in the case of all mathematical models, this is no doubt an oversimplification of reality. The business manager is usually most concerned with secular trend and seasonality. Informally, you can often identify these components just by looking at a time series plot. The function of the more formal time series analysis is to describe the components mathematically and to incorporate them into a mathematical model.

Decomposition of Time Series

Key Concept 53

Decomposition of a time series is a procedure for identifying and describing each of the four components of a time series.

Key Concept 54

Seasonal adjustment of a time series means removing the seasonal component.

The purpose of the decomposition of a time series is to see which of the four components described above are present and to describe them as precisely as possible. Sometimes one or more of these components may obscure the others and special techniques may be used to clarify the picture. It may be, perhaps, that you want to "remove" the cyclical and seasonal components from the picture in order to get a clearer picture of the secular trend. Removing the seasonal effect is called **seasonally adjusting** the data. If seasonal fluctuations can be clearly seen and accounted for, this will contribute to a simpler picture of influences not affected by seasonality, especially the secular trend and long-term cycles, which can then be projected ahead with more confidence. For example, the seasonal rise in retail unemployment just after Christmas may mask the fact that unemployment has generally been decreasing for a longer period of time. In contrast to this, the manager who wants to forecast sales for next March needs to include the seasonal component as well as the trend.

Alternatively, in the field of economics, the determination and description of cyclical patterns is very important, and the analysis may be slanted towards the hunt for hidden or obscure periodicities. In this case it is very helpful to be able to remove both seasonal variation and trend in order to see cycles more clearly. What you do depends on what you want. For the purposes of this text, I will concentrate on describing trend and seasonal component because they are frequently of concern in business applications, and because there are some fairly simple ways to approach the problem.

Table 8-2. Fake Time Series with Sawtooth Secular Trend and Cycle of Length Three

Time	Value	Time	Value	Time	Value
1	94	6	115	11	79
2	100	7	94	12	73
3	94	8	100	13	94
4	115	9	94	14	100
5	121	10	73	15	94

Smoothing a Series— Moving Averages

Key Concept 55

Smoothing a time series is a technique for revealing underlying patterns. Two elementary smoothing techniques are the moving average and exponential smoothing.

One problem which occurs frequently in the analysis of time series is the appearance of irregularities or random deviations from the presumed underlying smooth pattern. A useful tool for "removing" these deviations and "smoothing" a time series is the **moving average**. A moving average series is a new, less jagged series derived from an original, more irregular one, but which retains the basic long-term trend of the original. It is easiest to see how this technique works with a small fake data set.

Example 8-1: Fake Series. This fake time series was constructed by first assuming a sawtooth-like underlying trend. Then data points were placed near the sawtooth in sets of three, where the first and third points have the same value and the middle one is higher. The made-up time series of 15 values is given in Table 8-2 and graphed in Figure 8-2.

The series as presented is somewhat jagged. We attempt to smooth it by constructing a moving average series from it. The first step is to decide on the period or length of the cycle that seems to be repeating on top of the underlying trend. You can decide this yourself, for example, by looking at the distance between peaks, which is three in this graph. Or, you may have a reason based on some principle for knowing the period, which is usually the case when you assume a one year period. For this particular series, the period is three because that is the way the numbers were made up. Within each unit of three values which make up a cycle, the same pattern of ups and downs occurs. It therefore follows that if each set of three consecutive values is replaced by its own average, the ups and downs which occur within that cycle will be eliminated, or "smoothed out."

To start, find the mean of the first three values. Mean 1 equals $(94 + 100 + 94)/3 = 96$. Then we find mean 2 by *moving* over one value and taking the mean of the three values beginning there. So, mean 2 equals $(100 + 94 + 115)/3 = 103$. Continue moving this way as far as possible, listing the mean next to the *middle number* of its set of three. The results are in Table 8-3. Then both the original time series and the moving average series are plotted together in Figure 8-3.

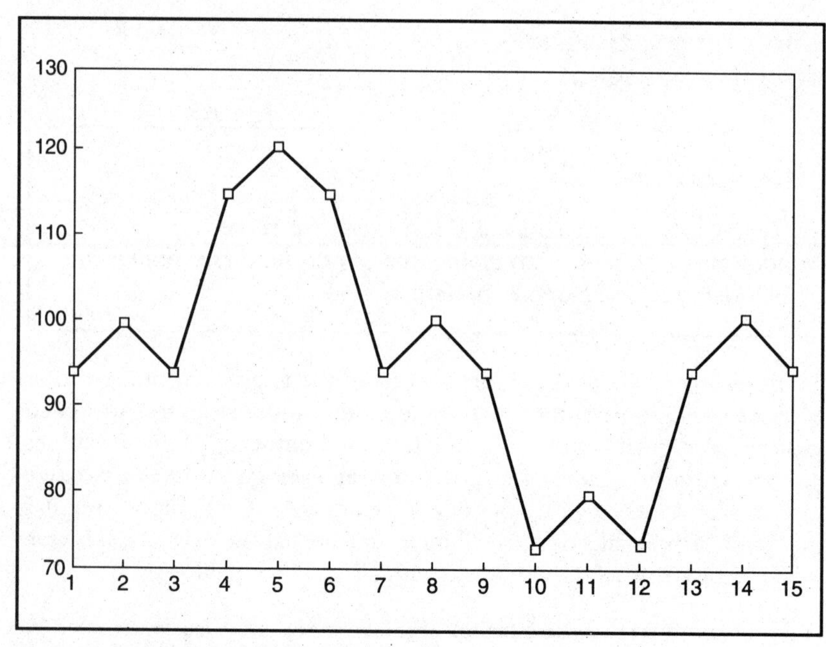

Figure 8-2. Time series graph based on Table 8-2.

Table 8-3. Moving Average Series, Period of Length 3, Based on the Series in Table 8-2

Time	Data values	Moving average
1	94	
2	100	96
3	94	103
4	115	110
5	121	117
6	115	110
7	94	103
8	100	96
9	94	89
10	73	82
11	79	75
12	73	82
13	94	89
14	100	96
15	94	

From the sketch it is evident that the moving average series does reveal the underlying sawtooth. In general, a moving average does the best job of showing an underlying trend when the number of values used to compute each mean matches the length of the cycle. If you use too few values for each mean, then the result may be too jagged; if you use too many, the result may be too flat. *For a time series made up of monthly recorded values, the smoothing is done by using averages of 12 consecutive values* because there is some yearly pattern which starts over every 12 months. Similarly, a *time series of quarterly values would be smoothed by averages of 4 consecutive values.* In contrast to simple regression, the shape of the underlying pattern is not specified in advance. (Recall that in simple regression we *assume* the underlying pattern is linear.)

Once you have a picture of an underlying pattern you can then describe more carefully how each point of the series departs from that pattern. For a series with a seasonal component, this is done by computing a seasonal *index*, which we describe a little later in more detail.

Case Study—Electricity Sold by Rural Utility: Monthly Time Series

In early 1991, as part of an application for a new rate structure, a supplier of electricity in Central Vermont wanted to describe the pattern of electricity usage among its residential and farm customers

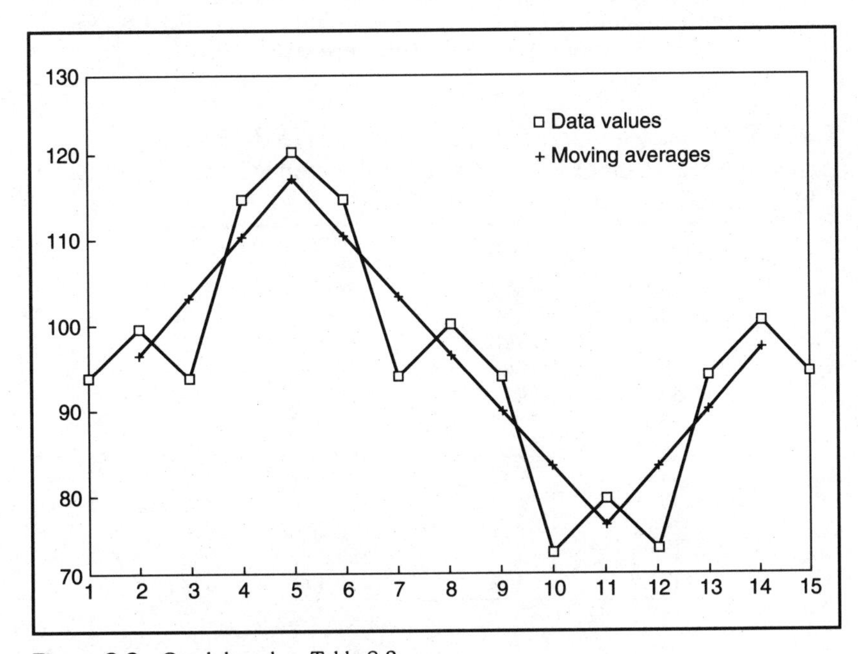

Figure 8-3. Graph based on Table 8-3.

over the previous several years. The data to be used were the monthly KWH billed by the electric cooperative to its residential and farm customers for a four-year period from January 1987 to December 1990. These data do comprise a time series because each observation is associated with a specific time, and the times are regularly spaced.

Step 1. Look at the data.

The data are presented in Table 8-4 and plotted in Figure 8-4.

From the wave-like pattern it can be seen that the series exhibits some regular variation. Based on the relationship of energy use and weather, it makes sense to assume that the cycle repeats every year—that it is a 12-month cycle. This can also be seen in the data because the distance from one peak to the next is about 12 months. One may then inquire about any general secular trend "underneath" the cycling back and forth. Is there a steady increase or decrease in energy use over this four-year period?

It is worth noting here that the KWH usage for December 1989 is unusually high. This spike was driven by an exceptional cold snap, which produced the coldest December on record for many years. Values which are noticeably more extreme than the others in a distribution are called **outliers** and they may have great impact on the calculations. For this utility company, any projections should take into account the probability of such cold snaps: perhaps a contingency fund should be put aside for associated expenses assuming (based on the given four years of data) that they occur no more frequently than once every four years.

Table 8-4. Kilowatt Hours of Electricity Billed to Residential and Farm Customers of a Rural Electric Cooperative from 1987 to 1990 (the units are 10,000 KWH)

Month	1987	1988	1989	1990
Jan	459.0	474.2	475.1	522.4
Feb	446.6	469.7	454.6	437.5
Mar	394.6	416.3	412.1	391.3
Apr	354.1	326.6	340.9	426.9
May	373.6	411.7	375.3	358.2
Jun	359.4	353.0	331.9	354.6
Jul	315.3	330.9	342.4	353.3
Aug	337.2	359.9	349.1	336.0
Sep	352.9	343.7	360.0	381.9
Oct	295.1	311.3	390.2	376.7
Nov	454.3	452.9	395.3	398.2
Dec	457.2	485.8	610.4	473.0

Step 2. Smooth the data by a moving average.
To get a clearer view of any trend underlying the monthly variations, we create a moving average series using 12 consecutive values for each mean. In the earlier section about moving averages, it was noted that each average is plotted (or tabled) next to the *middle* value of its set. For the first set of 12 values, the middle value position is halfway between month 1 and month 12, which is "month" 6.5. Then, the next mean goes at position 7.5, and so forth. This gives a list of 37 means as shown in column 4 of Table 8-5. In order to center the moving averages *on* months, rather than between them, the moving average process is *repeated* one more time using *two* consecutive values for each mean. This will center the first new value at position 7 (half-way between 6.5 and 7.5) where it can be taken to be an underlying typical figure for month 7. If this method is followed, we end up with the set of 36 values shown in column 5 of Table 8-5. This new series is called "centered." Its graph is shown in Figure 8-5.

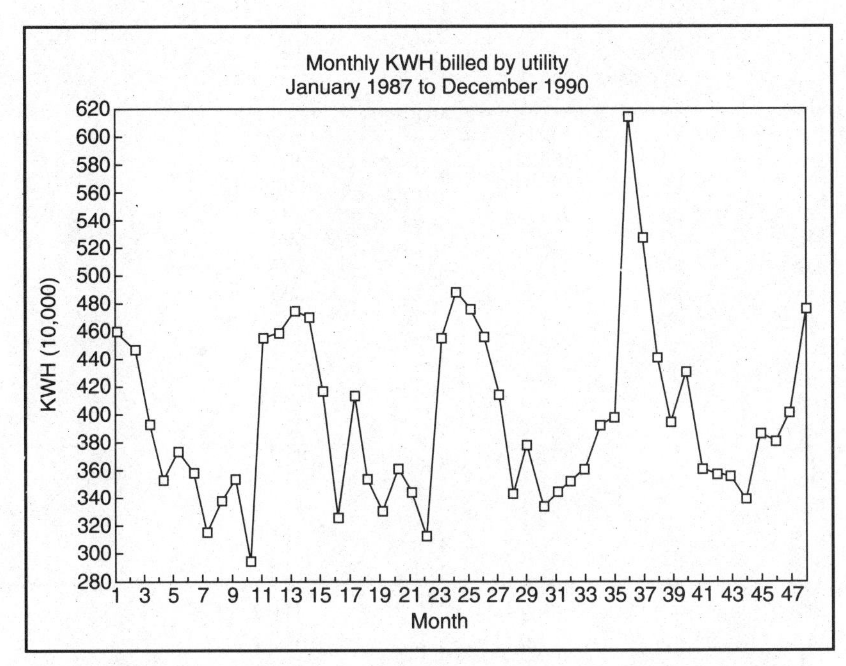

Figure 8-4. Time series plot of the data in Table 8-4. January 1987 is month 1, December 1990 is month 48.

Table 8-5. 12-Month Moving Averages for the Time Series in Table 8-4

Date	Month number	KWH	Moving average	Centered average
Jan 87	1	459.0		
Feb	2	446.6		
Mar	3	394.6		
Apr	4	354.1		
May	5	373.6		
Jun	6	359.4	383.3	
Jul	7	315.3	384.5	383.9
Aug	8	337.2	386.5	385.5
Sep	9	352.9	388.3	387.4
Oct	10	295.1	386.0	387.1
Nov	11	454.3	389.2	387.6
Dec	12	457.2	388·6	388.9
Jan 88	13	474.2	389.9	389.3
Feb	14	469.7	391.8	390.9
Mar	15	416.3	391.1	391.4
Apr	16	326.6	392.4	391.7
May	17	411.7	392.3	392.3
Jun	18	353.0	394.7	393.5
Jul	19	330.9	394.7	394.7
Aug	20	359.9	393.5	394.1
Sep	21	343.7	393.1	393.3
Oct	22	311.3	394.3	393.7
Nov	23	452.9	391.3	392.8
Dec	24	485.8	389.5	390.4
Jan 89	25	475.1	390.5	390.0
Feb	26	454.6	389.6	390.0
Mar	27	412.1	391.0	390.3
Apr	28	340.9	397.5	394.2
May	29	375.3	392.7	395.1
Jun	30	331.9	403.1	397.9
Jul	31	342.4	407.1	405.1
Aug	32	349.1	405.6	406.3
Sep	33	360.0	403.9	404.8
Oct	34	390.2	411.1	407.5
Nov	35	395.3	409.6	410.3
Dec	36	610.4	411.5	410.6
Jan 90	37	522.4	412.4	412.0
Feb	38	437.5	411.3	411.9
Mar	39	391.3	413.2	412.3
Apr	40	426.9	412.0	412.6
May	41	358.2	412.2	412.2
Jun	42	354.6	400.8	406.6
Jul	43	353.3		
Aug	44	336.0		
Sep	45	381.9		
Oct	46	376.7		
Nov	47	398.2		
Dec	48	473.0		

Describing the Underlying Trend

You might first note that these moving average values do not vary nearly as much as the original series. The seasonal fluctuations have been removed. The smoothed values range from a low value of 383.9 to a high value of 412.6 and very gradually increase as the months go on. Therefore, it is not unreasonable to *assume that the underlying secular trend is linear* with small positive slope. A formula for that line can be found by using simple regression on the original data set. Also, simple regression on the centered values could be used, and the resulting formula would not be very different; but it is more common to use all the original data.

Using the original data, the best-fitting line is given by $Y = 387 + 0.363X$, where Y is the KWH billed and X is the numerical value of the month from 1 to 48 in the original series. The line increases 0.363 units for each unit increase in X. This represents a steady increase of about 3.630 KWH billed per month. In Figure 8-5 both the centered moving average series and the regression line are superimposed on the original graph.

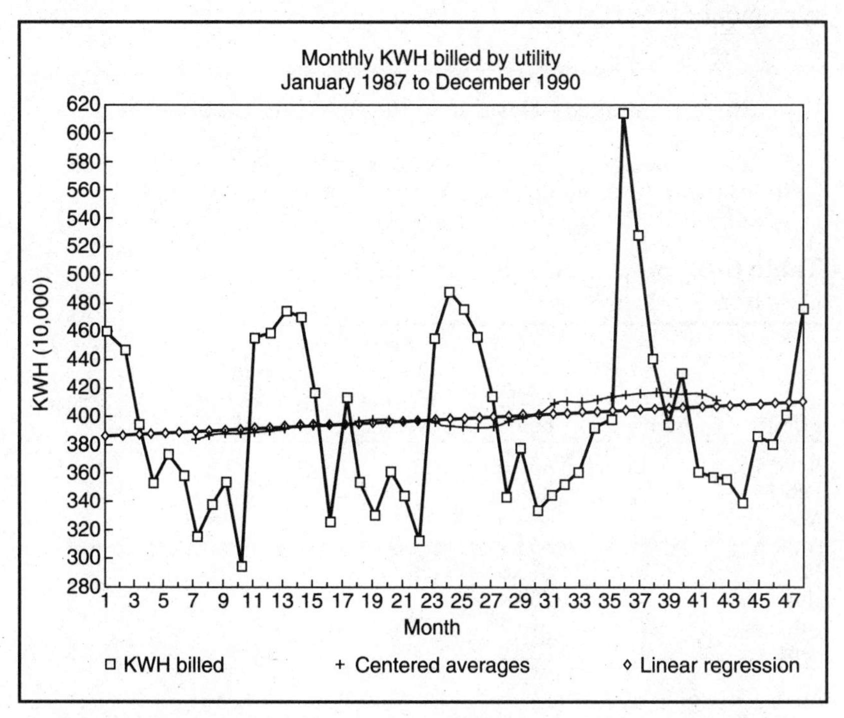

Figure 8-5. Centered 12-month moving averages and regression line superimposed on the time series from Figure 8-4.

Describing the Seasonal Component of a Time Series

For a monthly time series, you will probably want to describe the seasonal, or monthly, component. This is done by computing an index for each month which indicates how the value for the month typically compares to the underlying trend. One way to do this is to express the value for a specific month as a percentage of the 12-month centered average for that month. For instance, in Table 8-5, the KWH for July 1987 is 315.3 and the centered 12-month moving average for July 1987 is 383.9.

We divide the first figure by the second and multiply by 100 to get a specific index for July 1987.

$$\text{Index for July 1987} = \frac{315.3}{383.9} \times 100 = 82.1$$

This means that the KWH billed for July 1987 was 82.1 percent of the average billing for the 12-month stretch which has July 1987 in the center. In short, the KWH billed for July 1987 is low compared to the average amount billed.

Case Study Continued—Describing the Seasonal Component

Following the approach of the previous paragraph, the specific index for each month from July 1987 to June 1990 is given in Table 8-6.

Table 8-6. Specific Seasonal Indexes; Data from Table 8-5

Month	Month number	KWH billed	12-month centered moving average	Specific monthly index
Jan 87	1	459.0		
Feb	2	446.6		
Mar	3	394.6		
Apr	4	354.1		
May	5	373.6		
Jun	6	359.4		
Jul	7	315.3	383.9	82.1
Aug	8	337.2	385.5	87.5

Table 8-6. Specific Seasonal Indexes; Data from Table 8-5 *(Continued)*

Month	Month number	KWH billed	12-month centered moving average	Specific monthly index
Sep	9	352.9	387.4	91.1
Oct	10	295.1	387.1	76.2
Nov	11	454.3	387.6	117.2
Dec	12	457.2	388.9	117.6
Jan 88	13	474.2	389.3	121.8
Feb	14	469.7	390.9	120.2
Mar	15	416.3	391.4	106.4
Apr	16	326.6	391.7	83.4
May	17	411.7	392.3	104.9
Jun	18	353.0	393.5	89.7
Jul	19	330.9	394.7	83.8
Aug	20	359.9	394.1	91.3
Sep	21	343.7	393.3	87.4
Oct	22	311.3	393.7	79.1
Nov	23	452.9	392.8	115.3
Dec	24	485.8	390.4	124.4
Jan 89	25	475.1	390.0	121.8
Feb	26	454.6	390.0	116.6
Mar	27	412.1	390.3	105.6
Apr	28	340.9	394.2	86.5
May	29	375.3	395.1	95.0
Jun	30	331.9	397.9	83.4
Jul	31	342.4	405.1	84.5
Aug	32	349.1	406.3	85.9
Sep	33	360.0	404.8	88.9
Oct	34	390.2	407.5	95.8
Nov	35	395.3	410.3	96.3
Dec	36	610.4	410.6	148.7
Jan 90	37	522.4	412.0	126.8
Feb	38	437.5	411.9	106.2
Mar	39	391.3	412.3	94.9
Apr	40	426.9	412.6	103.5
May	41	358.2	412.2	86.9
Jun	42	354.6	406.6	87.2
Jul	43	353.3		
Aug	44	336.0		
Sep	45	381.9		
Oct	46	376.7		
Nov	47	398.2		
Dec	48	473.0		

Table 8-7. Monthly Indexes for Data from Table 8-6

Month	Median index	Monthly index
Jan	121.8	122.7
Feb	116.6	117.4
Mar	105.6	106.3
Apr	86.5	87.1
May	95.0	95.7
Jun	87.2	87.8
Jul	83.8	84.4
Aug	87.5	88.1
Sep	88.9	89.5
Oct	79.1	79.7
Nov	115.3	116.1
Dec	124.4	125.3
Total	1191.7	1200.0
Mean	99.3	100

At this point each of 36 individual months has its own specific index. There is a different index for July 1987, July 1988 and July 1989 for example. The next step is to combine all the indexes which are for the same "season," that is, combine all the Julys, then all the Augusts, etc., to get a "typical" index for each month. The procedure now becomes rather ad hoc. You could use any reasonable average. It is common, in the interest of smoothing, to throw out the highest and lowest values and then take the mean of the remaining ones. It is also common to use the median index. For the data in Table 8-6, because there are three indexes for each month, the two methods are equivalent. These medians are shown in Table 8-7. For instance, the three January indexes are 121.8, 121.8, and 126.8, so the median January index is 121.8.

Note that the sum of all 12 medians is 1191.7 which makes their average 1191.7/12 = 99.3. It would be easier to interpret these individual indexes if their mean were 100, and so to accomplish this, a final adjustment is made to get what we will call "the" monthly index. Each of the 12 medians is multiplied by 1200/1191.7, which increases each one just enough to bring the sum to 1200 and the mean to 100. These adjusted indexes are printed in the last column of Table 8-7. It is this last set of monthly indexes which summarizes the degree to which an individual month tends to depart from the underlying trend. For example, the index for December is 125.3. This means that the KWH billed in December is typically about 25 percent above the trend line. In Table 8-7 and Figure 8-6 you can clearly see the pattern of increasing KWH billed in the cold months and decreasing KWH billed

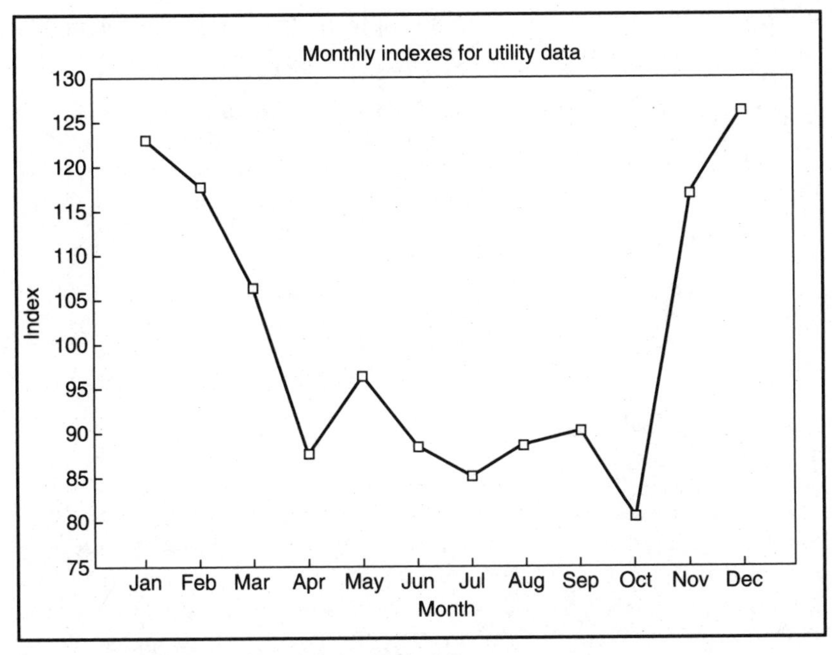

Figure 8-6. Monthly indexes listed in Table 8-7.

in the warm months. It is interesting to note the way that April and October don't quite fit the pattern; KWH billed in October is lower than a smooth pattern might suggest, as is April. Perhaps this is due to switching between Standard and Daylight Time.

Forecasting— Applying Seasonal Index to Secular Trend

Key Concept 56

To get the seasonal forecast multiply the underlying trend value by the seasonal index.

When you have a model which can be used to describe the underlying secular trend, then it can be adapted to forecast for a particular season. All that is necessary is to multiply the trend value by the appropriate seasonal index.

Example 8-2: Forecast the KWH Billed for January 1991. The secular trend was linear, represented by the formula KWH = 387 + 0.363 (month). January 1991 is the forty-ninth month in the series, so the trend value is

$$\text{KWH} = 387 + 0.363\ (49) = 404.8.$$

The seasonal index for January is 122.7, and so the forecast is 122.7 percent of 404.8 which is $1.22 \times 404.8 = 497$.

Seasonal Adjustment

One obvious reason for managers to be aware of seasonal fluctuations in demand is to make effective use of inventory. If you know the seasonal pattern in sales, for example, you can take best advantage of storage facilities and insurance costs. You can think about the risks involved with forced selling; you can try to avoid overproduction. You can plan reassignment of workers. But the need to think about seasonality clearly varies among various businesses; a farmer's production may be severely determined by seasonal fluctuations while a stock broker's sales may have almost nothing to do with the seasons.

Seasonal adjustment is a procedure for *removing* the seasonal influences from a series. As such it is another method for seeing underlying patterns. The simplest method of seasonal adjustment is just to divide the observed value by the appropriate seasonal index. More complex schemes, such as those used by government economists for analyzing national economic trends, allow for evolution of the seasonal indexes over time. For many government series, both the original data and the seasonally adjusted data are published.

To create the seasonally adjusted series, each original value is divided by its seasonal index (and multiplied by 100 for proper scale). As an example, for the utility data of Table 8-4, the original value for January 1987 is 459.0 KWH, and from Table 8-7 the seasonal index for January is 122.7. Therefore, the seasonally adjusted value for January 1987 is $(459.0/122.7) \times 100$, which equals 374.1 KWH. The original and seasonally adjusted series are presented in Table 8-8 and Figure 8-7.

Looking at the seasonally adjusted data shows again that the long-term trend for KWH demand has been relatively flat over the four-year period, but that in 1990 demand has been more erratic than seasonality can account for. This is due mainly to unusual weather patterns for the year.

Table 8-8. Original and Seasonally Adjusted Series; Utility Data from Table 8-4

Date	Month	KWH	Seasonally adjusted KWH	Date	Month	KWH	Seasonally adjusted KWH
Jan 87	1	459.0	374.1	Jan 89	25	475.1	387.2
Feb	2	446.6	380.4	Feb	26	454.6	387.2
Mar	3	394.6	371.2	Mar	27	412.1	387.7
Apr	4	354.1	406.5	Apr	28	340.9	391.4
May	5	373.6	390.4	May	29	375.3	392.2
Jun	6	359.4	409.3	Jun	30	331.9	378.0
Jul	7	315.3	373.6	Jul	31	342.4	405.7
Aug	8	337.2	382.7	Aug	32	349.1	396.3
Sep	9	352.9	394.3	Sep	33	360.0	402.2
Oct	10	295.1	370.3	Oct	34	390.2	489.6
Nov	11	454.3	391.3	Nov	35	395.3	340.5
Dec	12	457.2	364.9	Dec	36	610.4	487.2
Jan 88	13	474.2	386.5	Jan 90	37	522.4	425.8
Feb	14	469.7	400.1	Feb	38	437.5	372.7
Mar	15	416.3	391.6	Mar	39	391.3	368.1
Apr	16	326.6	375.0	Apr	40	426.9	490.1
May	17	411.7	430.2	May	41	358.2	374.3
Jun	18	353.0	402.1	Jun	42	354.6	403.9
Jul	19	330.9	392.1	Jul	43	353.3	418.6
Aug	20	359.9	408.5	Aug	44	336.0	381.4
Sep	21	343.7	384.0	Sep	45	381.9	426.7
Oct	22	311.3	390.6	Oct	46	376.7	472.6
Nov	23	452.9	390.1	Nov	47	398.2	343.0
Dec	24	485.8	387.7	Dec	48	473.0	377.5

Simple Exponential Smoothing

Though we shall not go into detail on complex methods for analyzing time series, it is worthwhile to understand one technique which is a key part of such methods. Like the moving average approach previously described, simple exponential smoothing produces a new, less jagged series which mimics the behavior of the original one, and which can provide a basis for forecasting. This approach, though, forces a certain structure on the new series in order to reflect the assumption that each observation depends on the ones previous to it in a particular way. The relationship determined by simple exponential smoothing makes each observation depend most on the immediately preceeding one, less on the one before that, still less on the one twice removed, and so on, with

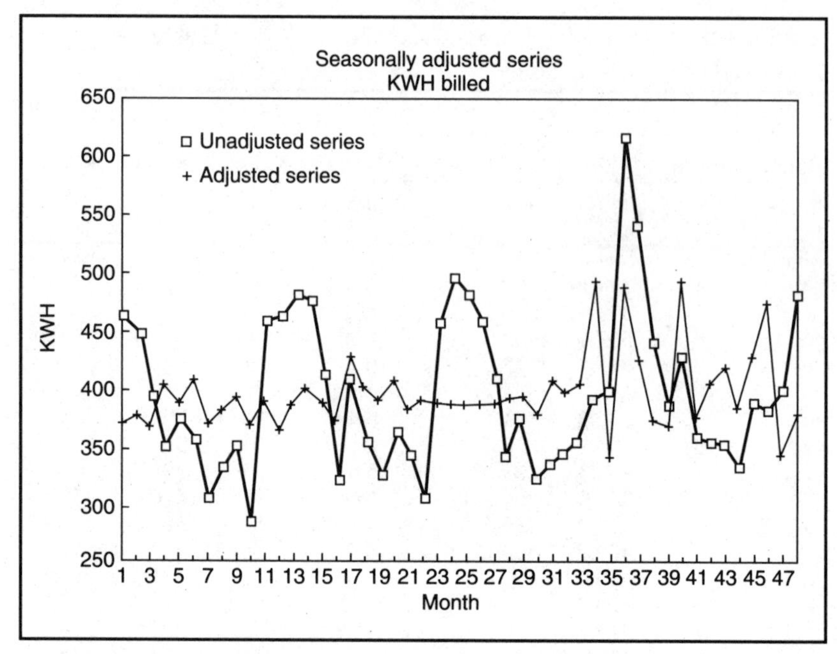

Figure 8-7. Adjusted and unadjusted series for utility KWH sales.

influence diminishing as the observations are further removed. Simple exponential smoothing, therefore, is a moving average approach which reflects a specific autocorrelation pattern. As such, it is a particularly simple example of an ARIMA model.

Simple exponential smoothing works best when the underlying trend is relatively flat, without pronounced trend or seasonality. If trend or seasonality is marked, then the technique can be modified to account for them. For example, one such approach, called the Holt-Winters model, provides estimates for these components and combines them to produce the new smoothed series. The Holt-Winters technique is often built into commercial forecasting software.

Table 8-9. Time Series

Time	Value
1	374
2	380
3	371
4	406

The result of simple exponential smoothing is to generate a new smoothed series in which each value can be expressed in terms of the previous ones. This is accomplished by making each of the smoothed values a weighted average of all the previous values, with recent values being given more weight than remote ones. To get the flavor of how this works, we apply the technique to a small time series, which is given in Table 8-9.

Example 8-3: Apply Simple Exponential Smoothing to a Time Series. The time series given in Table 8-9 consists of the first four values (rounded off) from the series in Table 8-4, the KWH billed by the utility company.

We start by picking a so-called **smoothing constant,** which is a value between 0 and 1. The lower the smoothing constant is, the less weight is given to observations further back in time. We will discuss later how to get a good value for this constant, but as a first example we arbitrarily use 0.4. The procedure becomes effective with the second number in the new series. The calculations are shown in Table 8-10 and explained below.

Each new entry in the smoothed series is found by multiplying the original value by (1–smoothing constant), multiplying the previous smoothed value by the smoothing constant, and taking the sum of these two calculations.

To see how this imposes the desired structure, consider the fourth value in the smoothed series, 393.0. It was calculated as:

$$393.0 = 0.6(406) + 0.4(373.6)$$

Note that the 393.0 contains the smoothed value 373.6 from one step back. The origin of 373.6 can be seen if replaced by its own formula.

$$\begin{aligned} 393.0 &= 0.6(406) + 0.4[0.6(371) + 0.4(377.6)] \\ &= 0.6(406) + 0.24(371) + 0.16(377.6) \end{aligned}$$

Table 8-10. Example of Simple Exponential Smoothing Where Smoothing Constant Equals 0.4; (1 – smoothing constant) = 0.6

Time	Original value	Smoothed value
1	374	374
2	380	0.6(380) + 0.4(374) = 377.6
3	371	0.6(371) + 0.4(377.6) = 373.6
4	406	0.6(406) + 0.4(373.6) = 393.0

So, the 393.0 actually depends on the smoothed value 377.6 from two steps back. In similar fashion, the origin of 377.6 can be seen.

$$393.0 = 0.6(406) + 0.24(371) + 0.16[0.6(380) + 0.4(374)]$$

$$= 0.6(406) + 0.24(371) + 0.096(380) + 0.064(374)$$

At this point, the fourth value is tracked back to its dependency on all the previous ones, where the weights decrease from 0.6 to 0.064 along the way. This is precisely the structure that exponential smoothing is intended to create. It is called "exponential" smoothing because each weight can be expressed in terms of powers of the smoothing constant.

$$0.6 = .4^0\ (.6)$$

$$0.24 = .4^1\ (.6)$$

$$0.096 = .4^2\ (.6)$$

$$0.064 = .4^3\ (.6)$$

It follows that the smaller the smoothing constant is, the more rapidly the remote values drop out of the picture.

Forecasting Using Exponential Smoothing

To use exponential smoothing as a forecasting device, you simply construct the new smoothed series and *use the last smoothed value as the forecast value for the next time period.* For the series in the previous example, the forecast for period 5 is 393.0. Then, when the true value is observed, you can continue smoothing to get the projection for the next period. In this sense, exponential smoothing is used to continually update forecasts as observations become available.

Determining the Smoothing Constant

The best smoothing constant is the one which produces a new series which mimics the original series as closely as possible. For a series which is very regular and consistent, with little variability, a constant close to 0 will work well, because the current forecast will not be very different from the immediately previous observation. It will not be important to retain the influence of observations long past. A constant near 0 allows the distant values to drop in influence rapidly. By contrast, the more irregular the series is, the closer to 1 the constant should be in order to retain the influence of distant values. It would be risky to let them drop out too soon because it is not true in an irregular series that each observation is just like the one which immediately precedes it.

A formal technique for determining the best constant is to measure the difference between the original and the smoothed series and then select the constant which minimizes this difference. This can be done in a straightforward way using the principle of least squares, by comparing each observation to its forecast value. In the next example we compare two choices for the smoothing constant, for the series in Table 8-8.

Example 8-4: Comparing Smoothing Constants. For the series of Table 8-8 compare the smoothing constants 0.4 and 0.8. Recall that the smoothed value at any given time serves as the forecast for the next time period.

Constant = 0.4

Time	Value	Smoothed value	Error	Squared error
1	374	374	374 – 380 = –6	36
2	380	377.6	377.6 – 371 = 6.6	43.56
3	371	373.6	373.6 – 406 = –32.4	1049.76
4	406	393.0		
			Total squared error =	1129.32

Constant = 0.8

Time	Value	Smoothed value	Error	Squared error
1	374	374	374 – 380 = 6	36
2	380	375.2	375.2 – 371 = 4.2	17.64
3	371	374.36	374.36 – 406 = –31.64	1001.09
4	406	380.688		
			Total squared error =	1037.11

Because the total squared error is smaller when the smoothing constant is 0.8, we would choose it over 0.4. The procedure for selecting the *best* smoothing constant between 0 and 1 is usually left for computer calculations.

Key Concept 57

Two popular statistics for evaluating how well a smoothed series mimics a given time series are the mean squared error (MSE) and the mean absolute percentage error (MAPE).

The value of the *mean* of the squared errors (MSE) is often used to summarize the "fit" of the smoothed series to the original series. In this case, for the smoothing constant equal to 0.8, we get MSE = 1037.11/3 = 345.7.

Sometimes analysts prefer to summarize the fit of a smoothed series in terms of percentage error, to reflect the percentage by which each smoothed value misses its corresponding actual value. One popular summary is to use the mean of all the individual percentage errors, ignoring their signs. This is called the MAPE, mean absolute percentage error. For the series above, with smoothing constant = 0.8, the MAPE is computed as follows.

Constant = 0.8

Time	Value	Smoothed value	Percentage error	Absolute percentage error
1	374	374	(374–380)/380 = –.016 = –1.6%	1.6
2	380	375.2	(375.2–371)/371 = .011 = 1.1%	1.1
3	371	374.36	(374.36–406)/406 = –.078 = –7.8%	7.8
4	406	380.688		

Total absolute percentage error = 10.5
MAPE = Mean absolute percentage error = 10.5/3 = 3.5

Self-Check

Answers to the following exercises can be found in Appendix A.

1. Which of these tables presents a time series? Draw a time series plot for it.

a.

Age group	Median income
25 and younger	$15,000
26 to 45	$30,000
46 to 59	$50,000
60 and older	$40,000

b.

Ethnic group	Median income
Black	$15,000
Hispanic	$30,000
Asian	$50,000
White	$40,000

c.

Year	Median income
1975	$15,000
1980	$30,000
1985	$50,000
1990	$40,000

2. Name the four components which are analyzed in the decomposition of a time series.

3. In analyzing a series of quarterly profit figures, a moving average is used to smooth the data. In the first step, how many observations go into the calculation of each average?
   a. 2
   b. 12
   c. 4

4. Which of these are methods for revealing underlying patterns in a monthly time series?
   a. seasonal adjustment
   b. 12-month moving average
   c. simple exponential smoothing

5. In a long series of monthly observations there is seen to be a rather consistent relationship between any observation and the one 12 months before it. This is an example of:
   a. autocorrelation
   b. linear regression
   c. trend

6. When an economist notes that the pattern of annual auto sales reflects the major depressions and upswings of the national economy, she is describing:
   a. seasonality
   b. secular trend
   c. cyclical behavior

7. From a monthly time series, the following specific indexes are computed for all the Februarys.

February:	1988	1989	1990	1991	1992
Index:	80.1	85.3	82.0	83.5	87.2

   Find "the" seasonal index for February by
   a. using the median.
   b. throwing out the highest and lowest values, then taking the mean.

8. If the value of the underlying trend for March is 380 and the seasonal index for March is 85.1, what is the forecast for March?

9. *a.* From these three specific indexes for July, compute a seasonal index for July.

July:	1990	1991	1992
Index:	112	115	111

   b. Suppose the value of the secular trend for July is $15,000. Use the answer from part *a* to make a July forecast.

10. In a monthly series the sum of the seasonal indexes is 1180. To adjust them so that the sum is 1200, all the indexes should be multiplied by what fraction?
   a. 1200/1180
   b. 1180/1200

11. For a monthly time series which runs from January 1989 to December 1991, the equation for the linear secular trend is $Y = 135 + 0.5X$, where X is the number of the time period, and Y is gross sales in thousands of dollars. What is the projected value of the trend for March 1992?

12. For the given data, compute the corresponding four-month *centered* moving average series.

	Quarter			
	1	2	3	4
1990	17	18	18	21
1991	22	24	20	19

13. In a monthly time series of regional housing starts, the seasonal index for October is 90. If the actual number of housing starts in October 1992 is 900, what is the seasonally adjusted number of starts for that month?
    a. 1000
    b. 810
    c. 990

14. *a.* Apply simple exponential smoothing to this time series using a smoothing constant of .2, and forecast the value for period 5.

Time	1	2	3	4
Value	100	104	96	102

    b. Compute the sum of squared errors for the smoothed series.
    c. Compute the MAPE for the smoothed series.

15. Apply simple exponential smoothing to the time series of Exercise 14, using a smoothing constant of 1. What happens? What happens if the smoothing constant is 0?

9
Index Numbers

Key Concepts

58. An index is used to track the progress of variables over time.
59. Indexes are useful for comparing two time series.
60. Indexes are useful for combining information from several variables.
61. A price index is one with fixed values for "quantity."
62. The Consumer Price Index may be used to account for inflation.

The Language of Statistics

Index number
Percentage change
Aggregate index
Value index
Price index
Consumer Price Index (CPI)
Adjusting for inflation
Quantity index

Creating and Interpreting Indexes

Key Concept 58

An index is used to track the progress of variables over time.

When wages are subject to cost-of-living adjustments through so-called COLA clauses in contracts, reference is usually made to the Consumer Price Index, which is intended to track the effects of inflation on buying power. In general, an **index** is a number intended to measure the performance of some variable or combination of variables over time. One time period is chosen as the base or reference period at which the index is arbitrarily set to 100. Then the index is calculated at other times, allowing for easy comparisons.

For instance, if in one year an index changes from 100 to 110, this represents a 10 percent increase. Similarly, if the index falls from 100 to 90 over two years, this represents a 10 percent decrease in two years, or an *average* decrease of 5 percent per year.

Example 9-1: Creating a Simple Index. The figures in Table 9-1 represent the annual cost of electricity over five years for some business. The series can be expressed as an index, where the base period is year 1.

Because year 1 is the base period and the cost of electricity in year 1 was \$1800, each value is divided by \$1,800 and multiplied by 100 to get the corresponding index number. See Table 9-2.

Note that all the index numbers are interpreted with respect to year 1. You can read, then, that the cost of electricity in year 5 was 10.6 percent higher than it was in year 1.

Table 9-1.
Cost of Electricity

Year	Cost (\$)
1	1,800
2	1,948
3	2,009
4	2,027
5	1,991

Table 9-2.
Electricity Cost Index

Year	Cost ($)	Index number
1	1,800	100.0
2	1,948	108.2
3	2,009	111.6
4	2,027	112.6
5	1,991	110.6

A different calculation is needed, however, to find the *percentage change from one year to the next.* For that, it is necessary to divide the difference in the two *consecutive* index numbers by the *earlier* of the two values. For instance, the percentage change from year 2 to year 3 is calculated as

$$\frac{(111.6 - 108.2)}{108.2} \times 100 = 3.1\%.$$

The results of these calculations are shown in Table 9-3. Here you see that the cost of electricity in year 5 was 1.8 percent lower than it was in year 4.

Key Concept 59

Indexes are useful for comparing two time series.

Indexes may be used to compare the performance over time of two or more variables. By using indexes you eliminate any differences of scale, such as when one variable is measured in tons and the other in gallons, or when one is reported in millions of dollars and the other in thousands of dollars. The index puts them on the same scale and makes direct comparison possible in terms of percentages.

Table 9-3. Annual Percentage Change in Electricity Cost

Year	Cost ($)	Index number	Percentage change in index number from previous year
1	1,800	100.0	
2	1,948	108.2	8.2
3	2,009	111.6	3.1
4	2,027	112.6	0.9
5	1,991	110.6	−1.8

Example 9-2: Comparison of Two Price Indexes. Table 9-4 is taken from *Business Statistics, 1986,* published by the United States Department of Commerce.

Table 9-4. Automobile Price Indexes

Year	New car prices	Used car prices
1967	100.0	100.0
1968	102.8	103.6
1969	104.4	103.1
1970	107.6	104.3

From the data, one can see that over this five year period, prices of both new and used cars increased in somewhat similar fashion, although used car prices took a slight dip in 1969. Note that you can use this data to track the behavior of the prices relative to 1967, but that you cannot discern the actual dollar amounts. The fact that the used car *index* was higher than the new car index in 1968 certainly does not mean that the average price for a used car was higher than the average price of a new car, but does mean that the *rate* of price increase from 1967 to 1968 was greater for used cars.

Key Concept 60

Indexes are useful for combining information from several variables.

An index is especially helpful when you want to track a combination of several components. An index which does this is called an **aggregate index**. As a simple example, suppose your company sells two products: a vitalizer kit (sold by the unit), and an energizer food supplement (sold by the pound). To develop an index which includes sales of both items, first find the dollar value of total sales in each year as shown in Table 9-5.

The total sales changed from $67,500 to $137,200 in one year. To show this as an index with year 1 as the base period, divide both total sales figures by the year 1 figure and multiply by 100. The results are shown in Table 9-6.

Table 9-5. Aggregate Sales

Item	Year 1			Year 2		
	Number sold	Unit price	Total sales	Number sold	Unit price	Total sales
Vitalizer kit	1050 kits	\$50/kit	\$52,500	2000 kits	\$51/kit	\$102,000
Energizer	1500 lb	\$10/lb	\$15,000	3200 lb	\$11/lb	\$ 35,200
			\$67,500			\$137,200

This index allows you to see that total sales have increased 103.3 percent. Note especially in this example that *both* the quantity sold and the price charged varied over time. Such an index is called a **value index**. As a result, the index conveniently summarizes the total change in the value of sales, but no longer shows the changes in the individual components. It would not be evident, for instance, from the index in Table 9-6, whether a change in the index was due mainly to changes in the quantities sold or to changes in prices.

Price Indexes

> **Key Concept 61**
>
> *A price index is one with fixed values for "quantity."*

A very common application of an index is to keep track of inflation or purchasing power. In such an index the *quantities are usually held fixed,* and only the changes in costs are reflected. This characterizes a **price index**. As an example, consider the **Consumer Price Index** (CPI), which is intended to track the price of the *same* "market-basket" of goods and services. Though, specifically, the CPI is an indication of the cost of only

Table 9-6. Aggregate Index

	Total sales	Index
Year 1	\$ 67,500	100
Year 2	\$137,200	203.3

these particular goods and services, it is broadly interpreted as describing the "cost of living," which is a much slipperier concept to quantify.

The base period which is currently used by the United States Department of Commerce for the CPI is the period from 1982 to 1984. (The base period is changed occasionally, so that it is not too distant in the past.) There are two somewhat different CPIs: CPI-W, based on the population of urban *w*age earners, and CPI-U, a broader based index based on all *u*rban consumers. As an illustration, the value of the CPI-U for July 1991 was 136.2, which implies that the more than 400 standard goods and services "that people buy for everyday living" cost 36.2 percent more in July 1991 than they did in 1982. The values of the CPI, and other more specialized price indexes (including farm and manufacturing indexes), are published monthly in the *Survey of Current Business* by the United States Department of Commerce.

Key Concept 62

The Consumer Price Index may be used to account for inflation.

Consider the case of a money time series covering several years, such as the value of an investment. It would be common that the value of the investment increases over time. But, even though the value has been increasing, one might feel that inflation should be taken into account to

Table 9-7. Deflation of a Time Series

Year	CPI-U (1982=100)	Value of investment at end of year (current dollars)	Value of investment at end of year (1982 dollars)
1985	107.6	$110,000	$102,230*
1986	109.6	$121,000	$110,401
1987	113.6	$133,100	$117,165
1988	118.3	$146,410	$123,762
1989	124.0	$161,051	$129,880
1990	130.7	$177,156	$135,544

*You get this by dividing 110,000 by 107.6 and then multiplying by 100.

describe more accurately the buying power of this investment. To achieve this, you can divide each value of the original series by the corresponding value of some appropriate price index. This then gives you the series expressed in "constant dollars." For example, if the base period were 1982, then all the new values would be expressed in "1982 dollars." It is not unusual to see that an increasing series may tell a quite different story after adjustment for inflation.

Example 9-3: Deflating a Time Series. Suppose that you invested $100,000 at 10 percent interest on January 1, 1985. Show its value through the end of 1990, and show the purchasing power of the investment according to the CPI (1982 = 100). The result is shown in Table 9-7, where the investment value is divided by the CPI. You can see that the nominal value of the investment grows by 10 percent annually, but the increase in purchasing power is substantially smaller.

In contrast to a price index, one might desire only to compare quantities. The increase or decrease in the number of units sold, or the number of customers serviced, may be of interest independent of any dollar figures. An index designed for this purpose is, therefore, called a **quantity index.**

An illustration of a quantity index in Table 9-8 is the reproduction of the index for auto and truck *production,* as given in the August 1991 *Survey of Current Business,* published by the United States Department of Commerce. For this index 1987 = 100.

This index allows you to see how the *number* of cars and trucks produced compares to what it was in 1987. In all cases, since the index is less than 100, it is clear that production is lower in 1991 than in 1987.

Table 9-8. Auto and Truck Production Index (1987=100)

Month (1991)	Index
Jan	79.6
Feb	74.7
Mar	76.7
Apr	85.0
May	89.2
Jun	92.5
Jul	98.1

Self-Check

The answers to the self-check exercises can be found in Appendix A.

1. An index often measures the performance of some variable over ____.

2. The base period for an index of apple prices was chosen to be 1990.
   a. What was the value of the index in 1990?
   b. In 1992 the index stood at 106. Interpret the 1992 value.

3. An index increases from 80 to 90 over a period of time. What was the percentage increase in the index between those two times?

4. The table shows the price of gold and silver (dollars per ounce) for certain months in 1991.

	Jan	Feb	Mar	Apr	May	Jun	Jul
Gold	383.64	363.83	363.34	358.39	356.82	366.72	367.51
Silver	4.028	3.723	3.960	3.970	4.040	4.390	4.300

(SOURCE: *Survey of Current Business,* October 1991)

   a. Convert both series into indexes using January as the base period.
   b. Draw time series graphs for the gold and the silver index. Draw both graphs on the same set of axes.
   c. What was the percentage change in gold between January and July?
   d. What was the percentage change in silver between January and July?
   e. Over the period January to July what was the average monthly percent change in the price of silver?

5. Refer to Table 9-5. Suppose that the data for year 3 are as follows.

	Number sold	Unit price
Vitalizer kit	2020	$53
Energizer	3300	$12

   a. Calculate the total sales for year 3.
   b. Compute the index number for year 3.

6. In a price index, which of these is kept fixed: price or quantity?

7. The table below shows the average price in dollars per gallon of home heating oil over several years.

Year	1985	1986	1987	1988	1989	1990
Price	.85	.56	.58	.54	.82	.99

   a. Use the CPI figures from Table 9-7 to "deflate" these prices and express them all in constant 1982 dollars.
   b. According to the calculations in *a,* is it correct to say that oil was "more affordable" in the years 1986 to 1990 than it was in 1985?

Appendix A
Answer Key to Chapter Self-Checks

Chapter 1

1. *a.* Descriptive statistics; no attempt is being made to generalize from these data.
2. *b.* Inferential statistics; the interpretation refers to the *larger* group which the sample represents.
3. *a.* The sample; they *represent* the population, which is *all* credit card holders.
4. *a.* Margin of error.
5. Lower bound = 10% – 3% = 7%.
   Upper bound = 10% + 3% = 13%.
6. Inferential statistics.
7. It means that he sold twice as many cars this month. Since the *number* of cars sold last month is not given, it is impossible to evaluate the significance of the increase. It could be as little as an increase from one to two cars sold.

Chapter 2

1. "Recent purchase" and "store credit card" are categorical, since the only responses are "yes" and "no." "Purchase amount" is a numerical variable.

2.
```
2 | 9
3 | 002234455566667777778889
4 | 001112223334445555556667899
```

3.

Interval	Frequency
27–29	1
30–32	4
33–35	6
36–38	12
39–41	6
42–44	9
45–47	9
48–50	3

4.

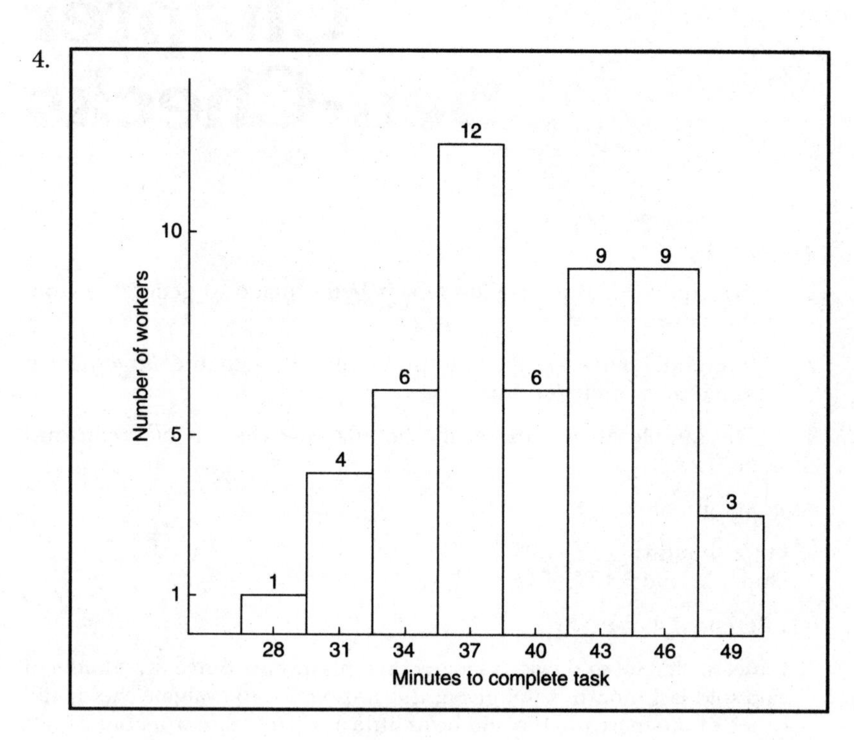

5. *a.* $n = 5 + 16 + 17 + 11 + 1 = 50$

b. 11 + 1 = 12 tangerines. $\frac{12}{50} = 0.24 = 24\%$.

c. The interval width is .20 ounces, because the difference between successive boundaries is .20.

6. A bar graph similar to the one in Figure 2-9 would work well. It could look something like this.

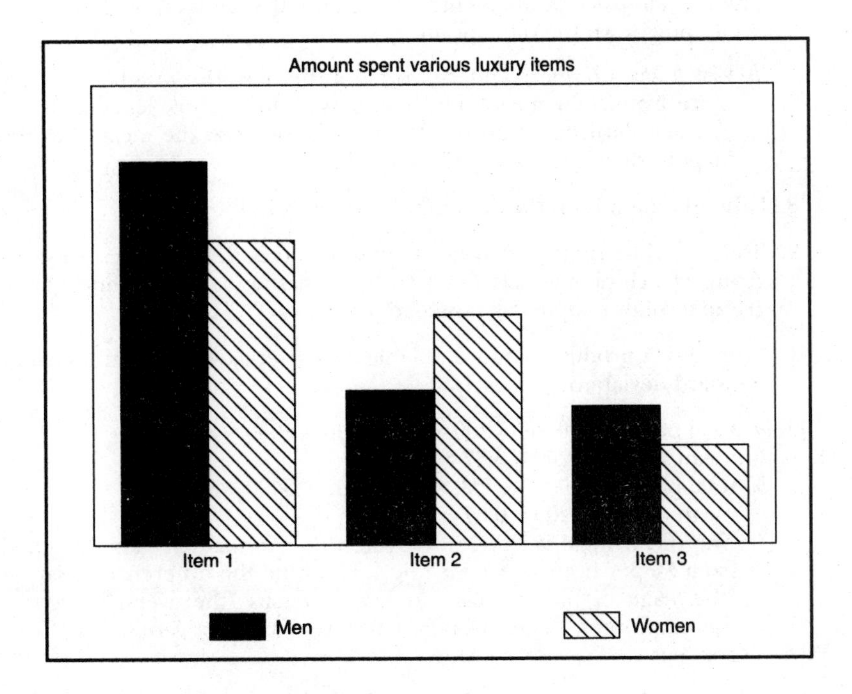

7. Increases would look less dramatic. For instance, a change from 0 to 20 would rise one inch in the first graph but only one-half inch in the second.

Chapter 3

1. *b.* The median, because the mean will be pulled towards the longer tail.
2. As a general statement this is false. In particular, it is false if the distribution of data is skewed; it is true if the distribution is symmetrical.
3. All three comments are correct.
4. *a.* It would be useful to know both the mean and the median.
5. The mode would be of more use. Because football numerals are used as "names" of players, they are categorical variables, and the mean or median

would not make any particular sense. The mode would indicate which numeral was sold the most, and could help in maintaining inventory.

6. The median price would be higher, because the tail is on the left, which would pull down the value of the mean.

7. *b.* List *b* has a larger standard deviation because the numbers deviate more from their mean. The numbers in list *a* show less variability. The standard deviation measures deviation *from* the mean, but the magnitude of the mean itself is not important.

8. False. It depends on the amount of variability in the data set.

9. Both could be right. For instance, they would both be correct if a small group of rich people had a very large increase in income, while a large group of poor people had a small decrease.

10. Company A produces a more reliable dispenser as shown by the smaller standard deviation.

11. *a.* total paid $= 1000 \times 6 + 200 \times 10 = 6000 + 2000 = 8000$.
    mean $= 8000/1200 = 6.67 = \$6.67$.
    b. total paid $= 1000 \times 5 + 200 \times 15 = 5000 + 3000 = 8000$.
    mean $= 8000/1200 = 6.67 = \$6.67$.
    The mean wage is still \$6.67. Note that the median wage dropped from \$6 per hour to \$5 per hour. Also note the difference in saying "the wage of the average employee" versus "the average wage of the employees." The "average employee" is not a very well defined concept.

12. Exam A: Pat's deviation is $115 - 100 = 15$ for a standardized score of $15/15 = 1$.
    Exam B: Pat's deviation is $425 - 400 = 25$ for a standardized score of $25/50 = 0.5$.
    Relative to the other employees, Pat did better on Exam A because the standardized score was higher.

13. The line was considered out of control at 1 P.M. At 1 P.M. the output fell below 360 pieces, which is two standard deviations ($2 \times 5 = 10$) below 370 pieces.

14. Thirty-six hours is one standard deviation below the mean. From the sketch of the normal curve shown in Figure 3-3, it can be seen that 16 percent of the observations in a normal distribution are lower than that. This means that about 16 percent of the batteries have lifetimes shorter than 36 hours, and will therefore fail before the guaranteed time.

15. Weighted mean = .80 × 24,000 + .20 × 30,000 = 25,200. The mean salary for all employees is $25,200. It is closer to the mean salary of the female managers because most of the managers are women.

16. Yes. Because there are 50 observations, the median weight is halfway between the twenty-fifth and twenty-sixth weights. There are 21 observations in the first two categories (5 + 16), so the twenty-fifth and twenty-sixth must be in the third category, which has the given boundaries.

Chapter 4

1. *a.* This is empirical probability because it is based on actual previous occurrences.

2. *a.* 1/200, because you are one of the 200 entrants, and each entrant has an equal chance of winning.
   b. You would expect to win about one out of every 200 games, or once in 200 weeks, which is just short of four years.
   c. E = P (win) (Net value of win) + P (Lose) (Net value of loss).

$$= \frac{1}{200}(99) + \frac{199}{200}(-1)$$

$$= \frac{99}{200} - \frac{199}{200}$$

$$= -\frac{100}{200}$$

$$= -\frac{1}{2}$$

   Thus, on average you expect to lose one-half dollar per game. In short, in the long run the odds are against you. For every time that you gain 99 dollars, there are 199 times that you will lose 1 dollar.

3. 2 and –.35 are mathematically impossible because they are outside the range 0 to 1.

4. There are three chances to fail for every two chances to succeed, so two out of five chances (or 40 percent) represents the *probability* to succeed.

5. P (Succeed) = .90. Thus P (Failure) = .10. So, the odds in favor are .90/.10 or 9 to 1.

6. *a.* P (Malfunction) = P (Start then Fail then Fail) = .10 (.20) = .02 or 2 percent.
   The product has a 2 percent chance of malfunctioning.
   b. First, note that the complement to "either one fails" is "no failures."
   P (No failures) = P (Start then Good then Good)
   = .90 (.90)
   = .81 or 81 percent.
   Therefore, the probability of at least one failure is 19 percent (100 percent – 81 percent).

7. *b.* An event and its complement must account for all possible occurrences.

8. If Sally is the trouble shooter, there is an 80 percent chance that the problem will be solved. Sally has an 80 percent success rate.

9. *a.* P (Sale is made given Region A) = 20/200 = .10 = 10 percent.
   b. P (Sale is made given Region B) = 20/600 = .033 = 3.3 percent.
   c. P (Sale is made over both regions) = 40/800 = .05 = 5 percent.
   d. No, sales response does depend on the region. The response rate is higher in Region A.

10.

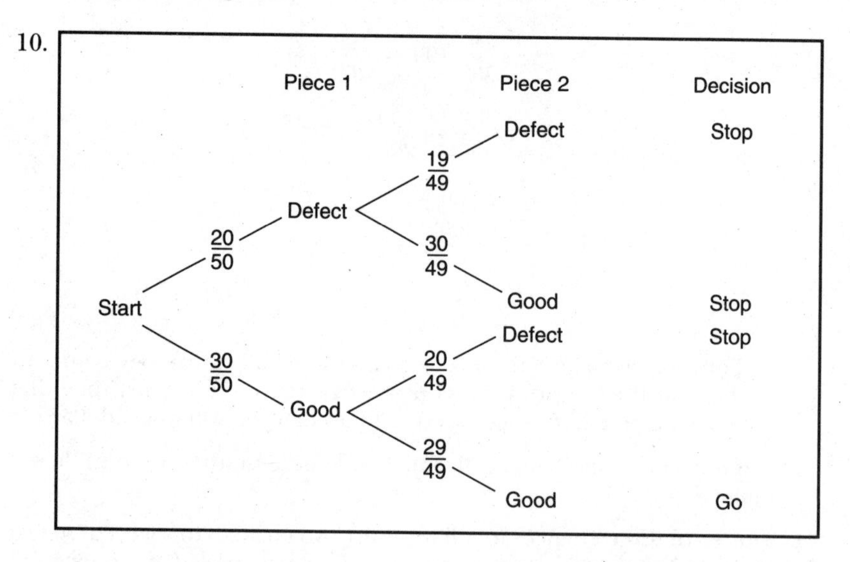

The event "line will be stopped" is the complement of "line will not be stopped."

P (Line will not be stopped) = P (Start then Good then Good)

= (30/50) (29/49) = .355

= 35.5 percent.

Thus, P (Line will be stopped) = 100 percent − 35.5 percent = 64.5 percent.

11. *a.* E = .60 (30) + .40 (45) = 18 + 18 = 36 days

    b. Ten projects are expected to take about 360 days, about 1 year.

12. *a.* Population: Unemployed women between the ages of 30 and 34 (as of June 1988).

    b. There were 904,000 such women.

    c. 47.7 per thousand of the 904,000 women had a baby during the year preceding June 1988. This is 47.7 × 904 = 43,120.8, or about 43 thousand babies.

13. The 10.1 birth rate is not interpretable because there are no units attached to it. Is this 10.1 births for every 100 people, or every 1000 people, or what?

Chapter 5

1. If you assume that the population of interest is all the people in Florida, then this is a parameter, because it characterizes the entire population.
2. This is a statistic because it is based on *sample* data.
3. Random.
4. Size.
5. Three digit numbers because 870, the largest number, has three digits.
6. 631, 858, 188, 491, ~~966~~, ~~997~~, 859. The values 966 and 997 are ignored because they are larger than 870.
7. *a.* It is called the *sampling distribution* because it consists of percentages collected in repeated sampling.
   b. The histogram will have the approximate shape of a normal curve.
8. Two.
9. *a.* The point estimate is 80 percent because 40 out of 50 is 80 percent.

   b. The standard error is $2\sqrt{\frac{.80(.20)}{50}}$

$$= 2\sqrt{.0032}$$

$$= 2\ (.0566)$$

$$= 0.113$$

$$= 11.3 \text{ percent.}$$

c. You can reduce the margin of error by increasing the sample size; you can investigate more than 50 mutual funds.

10. From Table 5-2 it can be seen that a sample of 1,111 loans will yield an estimate with a 3 percent margin of error.
11. The point estimate is 54.8 years because that is the mean of the sample.

The standard error is $\frac{s}{\sqrt{n}} = \frac{9.8}{\sqrt{100}} = \frac{9.8}{10} = 0.98$ years.

The margin of error is 2 (0.98) = 1.96 years. The estimated mean age of first class travelers is 54.8 years with a margin of error equal to 1.96 years.

12. This is a stratified random sample, because the various groups (strata) were deliberately included.

Chapter 6

1. Hypothesis.
2. Type I.
3. Null.
4. Samples. As long as the entire population is not examined, there is potential for error.
5. Test.
6. Statistically.
7. The test at the 5 percent significance level is more likely to result in a type I error. (5 percent chance versus 1 percent chance.)
8. Low.
9. You are hoping that the alternative hypothesis is true.
10. Sampling distribution.
11. Normal.
12. Contingency.
13. Two.
14. Rejection.
15. You would like the test statistic to fall in the rejection region.
16. More power.
17. Increase.

18. Yes, because the p-value is less than .05.
19. Type II.

Chapter 7

1. One. If there is more than one predictor, then you have multiple regression.
2. Dependent.
3. Model.
4. Coefficient.
5. Scattergram.
6.

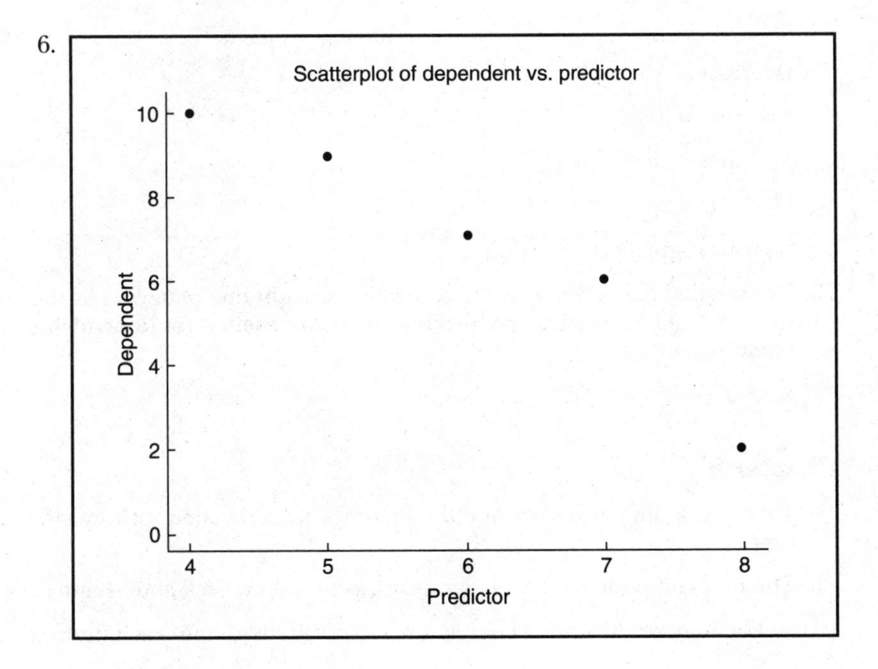

7. Increases.
8. A change of 1 unit in the predictor is associated with a change of 8 units in the dependent variable.
9. You would expect negative correlation because as unemployment increases, sales of new cars will decrease.

10. It is called the line of best fit, because it is the line that comes closest to fitting the pattern of the dots.

11. It fits better when r^2 is larger. Since .9 is larger than .86, it fits better when r^2 is .9.

12. The predictor variable accounts for 72 percent of the variability of the dependent variable. (More formally, the model accounts for 72 percent of the total sum of squares.)

13. The model yields expected sales = 2000 units.

    $E = 400 + 20\ (50) + 15\ (40) = 2000.$

    The margin of error is two times the standard error, which is 2 (10) = 20 units.

    So the 95 percent prediction interval is 2000 ± 20 units.

14. 200.

15. Decreases.

16. Are not.

17. Multiple.

18. Highest.

19. Logistic regression.

20. No, because they do not clearly indicate a straight line pattern. For this data, r^2 will be close to zero, making the data unsuited for dependable linear regression.

Chapter 8

1. Table *c* is a time series because the incomes are associated with specific times.

2. The four components are secular trend, seasonal, cyclical, and irregular.

3. *c.* For a quarterly series 4 consecutive observations go into each moving average.

4. All of these are methods for revealing underlying patterns.

5. *a.* Autocorrelation.

6. *c.* Cyclical.

7. *a.* The median of {80.1, 82.0, 83.5, 85.3, 87.2} is 83.5.
   b. The mean of {82.0, 83.5, 85.3} is 250.8/3 or 83.6.

8. Multiply the trend value by the seasonal index and divide by 100 for proper scale.

$$\frac{(380)\ (85.1)}{100} = 323.38.$$

This makes sense because the seasonal index of 85.1 indicates that the March value is typically about 85 percent of the secular trend. That is, March is a "low" month.

9. *a.* The index is the median, which is 112.
   b. Forecast is 112 percent of $15,000.

$$\frac{(15{,}000)\ (112)}{100} = \$16{,}800.$$

10. Multiply by *a.* 1200/1180.

11. March 1992 is the thirty-ninth month in the series, so take $X = 39$. This yields $Y = 135 + 0.5\ (39) = 154.5$.

12.

Time	Value	Four quarter moving average	Four quarter centered average
1	17		
2	18		
		18.5	
3	18		19.125
		19.75	
4	21		20.5
		21.25	
5	22		21.5
		21.75	
6	24		21.5
		21.25	
7	20		
8	19		

13. *a.* October is a "low" month because its seasonal index is 90. Thus, the seasonally adjusted number of housing starts for October (the secular

trend value) will be higher than the actual number of starts. Divide the actual number by the seasonal index and multiply by 100.

$(900/90) \times 100 = 1000$

14. *a.*

Time	Value	Smoothed value
1	100	100
2	104	.8 (104) + .2 (100) = 103.2
3	96	.8 (96) + .2 (103.2) = 97.4
4	102	.8 (102) + .2 (97.4) = 101.1

Forecast for time 5 is 101.1.

b. Sum of squared errors
$= (100 - 104)^2 + (103.2 - 96)^2 + (97.4 - 102)^2$
$= 16 + 51.84 + 21.16$
$= 89.$

c. MAPE is the Mean of the Absolute Percentage Errors.

Percentage error	Absolute Percentage error
$\frac{100 - 104}{104} = -0.038 = -3.8\%$	3.8%
$\frac{103.2 - 96}{96} = -0.075 = 7.5\%$	7.5%
$\frac{97.4 - 102}{102} = -0.45 = 4.5\%$	4.5%

$$\text{MAPE} = \frac{3.8 + 7.5 + 4.5}{3} = 5.3\%.$$

15. When the smoothing constant is 1, the first weight is 0 and the second weight is 1. This will make all the smoothed values equal to the first value (100). In short, the current value counts for nothing and the most remote past value counts for everything. When the smoothing constant is 0, the current value counts for everything and the past values count for nothing. The smoothed values are the same as the initial series; nothing changes.

Chapter 9

1. Time.

2. *a.* 100.
   b. The price of apples in 1992 was 6 percent higher than it was in 1990.

3. Percent increase is the difference between the values divided by the earlier value.

   $(90 - 80)/80 = 10/80 = 0.125$, which represents a 12.5 percent increase.

4. *a.* The indexes are computed by dividing each price by the January price and multiplying by 100.

Month	Gold price	Gold index	Silver price	Silver index
Jan	383.64	100.0	4.028	100.0
Feb	363.83	94.8	3.723	92.4
Mar	363.34	94.7	3.960	98.3
Apr	358.39	93.4	3.970	98.6
May	356.82	93.0	4.040	100.3
Jun	366.72	95.6	4.390	109.0
Jul	367.51	95.8	4.300	106.8

b.

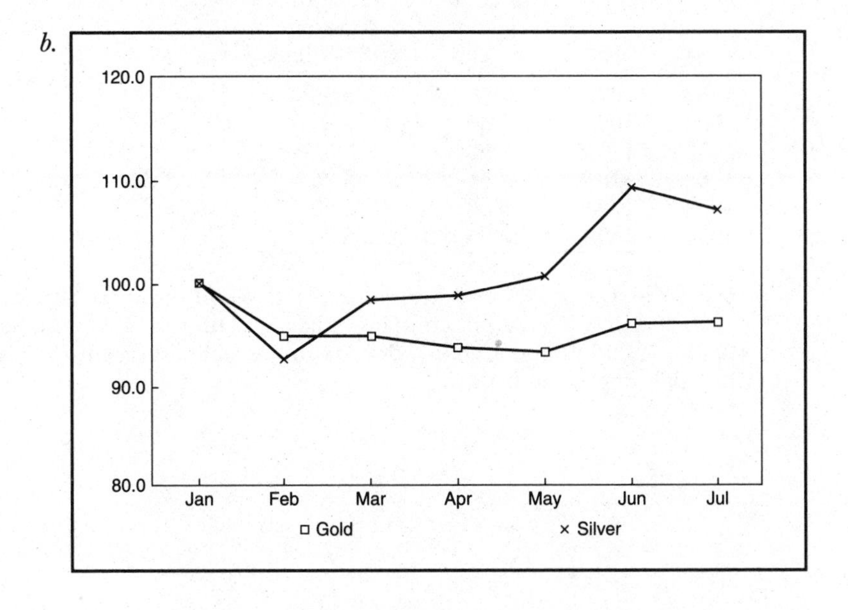

c. Since the index fell from 100 to 95.8 the change is $100 - 95.8 = 4.2$ percent. The price of gold was 4.2 percent lower in July than the previous January.

d. Since the silver index rose from 100 to 106.8, the change is $106.8 - 100 = 6.8$ percent. The price of silver was 6.8 percent higher in July than it was the previous January.

e. Silver rose 6.8 percent in six months for an average increase of 6.8/6 or 1.13 percent per month. Note that six months elapsed *between* the two dates.

5. *a.* Total sales for year 3
   = 2020 (53) + 3300 (12)
   = 146,660

   b. Get the index number for year 3 by dividing the total sales then by the total sales in year 1, $67,500.

   $$\text{Index number for year 3} = \frac{146{,}600}{67{,}500} \times 100 = 217.3$$

6. In a price index the quantities are kept fixed and the prices vary.

7. *a.* To compute the constant dollar figure divide the current dollar figure by the CPI value. Recall that for 1982 the index is 100; thus the results are all in 1982 dollars.

Year	CPI-U	Price in current dollars	Price in constant 1982 dollars
1985	107.6	$.85	$.80
1986	109.6	$.56	$.51
1987	113.6	$.58	$.51
1988	118.3	$.54	$.46
1989	124.0	$.82	$.66
1990	130.7	$.99	$.76

   b. It is correct to say that oil was more affordable in the years 1986 to 1990 because the price in constant dollars accounts for inflation. So "taking inflation into account" the 80 cent per gallon price in 1985 is the highest price in the list.

Appendix B
Formulas

1. **The standard deviation and variance for a set of data.** Symbol for standard deviation: s.

   s = square root of the mean of the squares of the deviations. The symbol s^2 represents the variance of the data.

   NOTE When the purpose of computing the standard deviation is to estimate the standard deviation of the population from which the data came, then $n - 1$ is used as the divisor instead of n in getting the "mean" of the squares. This computation provides an unbiased estimate for the variance. For large data sets it hardly matters whether the divisor is n or $n - 1$; the two formulas produce virtually the same answer.

   Example Find the standard deviation, s, for the data set 10, 12, 14, 16, 18.

   First, note that n is 5, $n - 1$ is 4, and the mean is 14. The computation depends on having the deviation of each value from the mean, which is found by subtracting the mean from each value.

Data	Deviation from mean	Square of deviation
10	−4	16
12	−2	4
14	0	0
16	2	4
18	4	16

Sum of squares of deviations = 40.

$$\text{"Mean" of squares of deviations} = \frac{40}{4} = 10.$$

$$\text{Standard deviation} = \sqrt{10} = 3.16.$$

In symbols, $s = 3.16$ and $s^2 = 10$.

2. **Formula for the normal curve**. This is the formula which mathematicians use to compute the percentage of area in various portions of the normal curve. In practice these areas are printed in tables or computed automatically by computer programs. The formula itself therefore is not used in ordinary applications of statistics.

$$y = \frac{1}{\sqrt{2\pi}} e^{-x^2/2}$$

3. **Formula for process control limits.** Calculations for Example 3-12. The limits are set 2 standard errors above and below the target value.

$$\text{Standard error} = \frac{s}{\sqrt{n}} = \frac{0.1}{\sqrt{25}} = \frac{0.1}{5} = 0.02 \text{ mm}$$

Upper limit: $75 + 2(0.02) = 75 + 0.04 = 75.04$

Lower limit: $75 - 2(0.02) = 75 - 0.04 = 74.96$

Appendix C
Tables

1. **Table of selected areas for the normal curve.** The amount of area in the table refers to the portion of the graph that is shaded in the sketch.

   The vertical lines are drawn x standard deviations away from the mean.

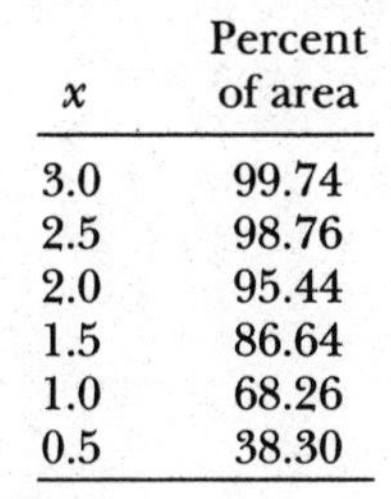

x	Percent of area
3.0	99.74
2.5	98.76
2.0	95.44
1.5	86.64
1.0	68.26
0.5	38.30

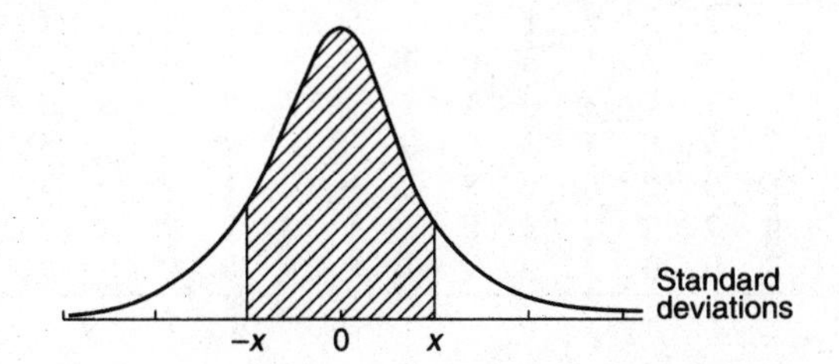

   In the discussions in this book we somewhat inexactly use $x = 2$ to enclose 95 percent of the area of a normal curve. In more exact calculations you would use $x = 1.96$. In practice, there is virtually no difference. Commercial software will use 1.96.

2. **Table of Random Digits.** This table was generated by computer. Similar lists can be generated by most commercial statistics programs. In addition, such lists are available printed in book form. Many calculators also have a RAN# key which will produce a series of random integers.

Table of Random Digits

8	9	8	6	3	1	8	5	8	1	8	8	4	9	1	9	6	6	9	9
7	8	5	9	5	3	7	1	7	9	6	9	3	1	9	3	2	3	3	7
9	4	3	2	4	5	6	3	4	9	4	0	3	2	1	9	2	3	6	2
7	3	1	2	1	7	5	3	4	1	7	9	0	5	3	3	3	0	0	9
0	2	1	2	8	7	2	6	6	6	5	5	4	2	1	5	5	9	1	9
1	4	6	4	5	8	5	5	1	6	4	0	7	7	4	9	9	7	2	5
6	5	4	9	7	4	9	9	4	0	8	5	6	0	5	1	3	4	6	1
2	0	1	8	9	0	9	8	7	0	7	5	2	8	5	6	6	4	5	7
0	7	9	2	0	6	5	4	4	6	9	9	4	9	0	2	3	2	1	9
8	1	6	0	3	3	3	2	8	8	3	9	6	7	6	9	1	6	7	7
0	9	5	0	7	9	3	5	2	3	9	7	2	7	2	8	3	3	0	4
1	3	7	3	9	8	4	8	6	0	7	6	2	3	7	9	8	8	0	9
7	3	5	8	7	1	5	8	2	1	7	4	2	7	1	6	5	1	7	3
1	2	0	9	2	8	7	6	1	2	2	1	0	9	9	9	1	9	9	1
4	9	7	0	5	7	6	5	4	9	6	7	2	3	1	2	1	8	8	3
9	1	7	1	9	7	4	8	4	5	5	1	3	3	5	2	5	3	4	0
7	6	6	8	3	1	1	3	8	2	9	2	5	6	9	3	9	3	8	5
7	3	9	9	9	1	9	9	5	8	8	6	3	9	2	0	2	7	3	6
9	4	5	2	9	2	7	2	7	6	9	4	2	8	1	7	9	3	7	7
7	8	5	4	7	9	6	9	7	8	9	5	5	0	6	1	3	0	0	3
2	5	0	8	7	1	2	7	8	7	1	5	3	4	0	9	2	1	2	5
7	5	1	6	1	7	3	7	0	9	6	3	8	4	3	9	3	7	5	2
6	0	4	6	1	3	5	9	0	3	2	9	1	1	9	2	5	4	5	6
4	5	9	6	1	9	1	0	5	7	4	6	7	2	1	8	9	0	6	2
1	5	3	1	8	0	1	8	3	2	4	8	2	0	8	3	3	7	6	8
4	1	7	3	0	1	2	4	3	2	5	6	2	4	4	2	1	3	9	6
7	5	6	4	3	9	9	8	8	4	5	4	3	0	8	7	8	3	3	7
3	9	2	6	2	8	2	9	6	1	2	0	7	2	4	5	3	6	4	7
9	7	8	2	3	0	1	5	0	2	1	9	1	8	0	0	1	7	8	4
8	9	0	3	9	4	1	8	4	6	2	8	4	0	5	4	6	2	9	5

Appendix D

Calculations for Hypothesis Tests

The general idea is that in each case there is a best statistic for testing the given null hypothesis. This statistic is computed from the sample data. Then the value of the statistic is appraised by referring to the appropriate well known distribution (such as normal or t or chi-square or F). If the sample statistic is numerically extreme, then this counts as strong evidence against the null hypothesis.

1. **Chi-square test** as illustrated in Example 6-4.

   Step 1. Construct the tables of observed and expected values.

   Observed Values:

	White	Black	Asian	Hispanic	Total
Used the product	24	60	36	18	138
Did not use it	80	100	60	62	302
Total	104	160	96	80	440

   Expected Values:
   The table of expected values has the same row and column totals as the table of observed values. The expected value in each cell is found by multiplying the row total for the cell times the column total for that cell and then dividing by the grand total. The table of expected values contains values which support the null hypothesis (that the two variables are statistically independent). In this example the table of

expected values shows that all the ethnic groups had the same proportion (138/440 = 31.4%) of people who used the product. In short, use of the product does not depend on ethnic group.

	White	Black	Asian	Hispanic	Total
Used the produce	32.6	50.2	30.1	25.1	138
Did not use it	71.4	109.8	65.9	54.9	302
Total	104	160	96	80	440

Step 2.

1. For each pair of corresponding cells find the square of the difference between the observed and expected values and divide by the expected value. In this example there will be 8 such calculations, the first of which is $(24 - 32.6)^2/32.6 = 2.3$.
2. Find the sum of all the results from the previous step. In this example the sum is

$$2.3 + 1.9 + 1.2 + 2.0 + 1.0 + 0.9 + 0.5 + 0.9\,,$$

which equals 10.7. This is the value of the test statistic.

Step 3. Compare the value of the test statistic to the appropriate value from a chi-square table.

Note To use a chi-square table properly you must know a value called the "degrees of freedom," which are not discussed in this book. For further explanation see a more advanced text. But for problems like this one, the degrees of freedom can be calculated by simply counting the number of cells in the table of observed values and then subtracting the number of cells along the bottom and the right edges. In Example 6-1, the degrees of freedom is 8 – 5, which is 3. You just count the interior cells for this, not the totals.

2. Test of the **difference of the means of two independent samples** as illustrated in Example 6-6. In this test the sampling distribution of the test statistic is the t-distribution. Such analyses are often referred to as "t-tests."

Step 1. Compute the difference of the two sample means. In this example we get .05 hours.

Step 2. Compute a weighted average of the two sample variances. The weight for each of the means is one less than the size of the sample from which it was computed.

In this example we get $\frac{19\ (.40)^2 + 19\ (.43)^2}{19 + 19} = .17$

Step 3. Multiply the result of Step 2 by $\left(\frac{1}{n_1} + \frac{1}{n_2}\right)$.

In this example we get .17 (1/20 + 1/20) = .017.

This is an estimate of the variance for the sampling distribution.

Step 4. Compute the square root of the previous result. This gives the standard deviation of the sampling distribution.

In this example $\sqrt{.017} = 0.13$.

Step 5. Divide the difference from Step 1 by the result in Step 4. This standardizes the sample statistic, so that it can be appraised by a standardized table.

In this example we get .05/0.13 = 0.38.

This is the test statistic.

Step 6. Refer to a table for the t-distribution, using for the degrees of freedom the sum of the weights from Step 2. In this example the degrees of freedom is 38.

3. Analysis of a **matched pairs design** by t-test, as illustrated in Example 6-8.

   Step 1. Calculate all the differences and then the mean of the differences. This is 0.36.

   Step 2. Compute the variance for the differences. This comes to 0.013.

   Step 3. Divide the variance by the number of pairs, and then find the square root. This comes to $\sqrt{0.013/5} = 0.051$, and represents the standard deviation for the sampling distribution.

   Step 4. Divide the mean from Step 1 by the result of Step 3 to standardize the sample statistic. This gives 0.36/.051 = 7.06. This is the test statistic, which has a t-distribution with degrees of freedom 1 less than the number of pairs.

Bibliography

Deming, W. Edwards. *Elementary Principles of the Statistical Control of Quality.* Tokyo: Nippon Kagaku Gijutso Remmei, 1950. [The ideas behind process quality control and quality circles. Also see his *Some Theory of Sampling,* Dover, 1966, which lays out the basic issues very clearly, particularly what to watch out for in the design of surveys.]

Hanke, J., and A. Reitsch. *Understanding Business Statistics.* Homewood, IL: Irwin, 1991. [A comprehensive, introductory college level statistics text. Has more detail on forecasting than the Newbold book, and is more oriented to computer exercises than the Newbold book.]

Naiman, A., R. Rosenfeld, G. Zirkel. *Understanding Statistics.* 3d ed. New York: McGraw-Hill, Inc., 1983. [A general lighthearted introduction to statistics at about the level of this text, but with many exercises.]

Newbold, P. *Statistics for Business and Economics.* 3d ed. Prentice Hall, 1991. [A standard college level textbook which covers the topics of this book in more technical detail. Has more detail on quality control than the Hanke and Reitsch book.]

Tanur, J., ed. *Statistics: A Guide to the Unknown.* 3d ed. Belmont, CA: Wadsworth and Brooks/Cole, 1989. [Presents a variety of applications of statistics, including accounting.]

Tufte, E. *Envisioning Information.* Chesire, CT: Graphics Press, 1990. [Similar to the other book by Tufte, but wider in scope.]

Tufte, E. *The Visual Display of Quantitative Information.* Chesire, CT: Graphics Press, 1983. [The premier reference on what makes graphs good or bad. A beautiful book.]

U.S. Department of Commerce. Bureau of Economic Analysis. *Survey of Current Business.* Washington, D.C. [Published monthly. Contains many current business series, such as the unemployment rate, and various price, wage, and productivity indexes, including the Consumer Price Index.]

U.S. Department of Labor. Bureau of Labor Statistics. *Monthly Labor Review.* Washington, D.C. [Published monthly. Focuses particularly on the labor outlook. Includes current data for various series and indexes such as unemployment, earnings, collective bargaining agreements, injury and illness data, productivity, and prices.]

U.S. Department of Commerce. Bureau of the Census. *Statistical Abstract of the U.S.* Washington, D.C. [Published annually. This is the "standard summary of statistics on the social, political, and economic organization of the United States." Contains over 1000 tables of summary data with references to the sources.]

Index

About the Author

Robert Rosenfeld holds graduate degrees from Harvard and Columbia Universities, and is professor of mathematics and statistics at Nassau Community College, where he has taught since 1966. He has also been a visiting lecturer in statistics at Adelphi University, the Columbia University Graduate School of Public Health, and the University of Vermont. The textbooks he has written include *Understanding Statistics,* published by McGraw-Hill, Inc. In addition to his academic work, he is a private consultant in applied statistics.

Final Examination

The McGraw-Hill 36-Hour Statistics Course

If you have completed your study of *The McGraw-Hill 36-Hour Statistics Course,* you should be prepared to take this final examination. The exam consists of 60 questions, which span the entire book.

Instructions

1. If you like, treat this as an "open-book" exam and consult the text while taking it. That approach will help to reinforce your learning and to correct any misconceptions. On the other hand, if you want to test what's "in your head," you may prefer to take the examination without reference to any textbook.

2. Answer each of the test questions on the answer sheet printed at the end of the exam. Do so by placing the letter which indicates the answer you choose in the appropriate blank.

3. All questions are multiple-choice, and you should select the answer that you think represents the *best* among the choices.

4. You must answer 42 questions correctly to have a passing grade of 70 percent. A passing grade entitles you to receive a *certificate of achievement.* This handsome certificate, suitable for framing, attests to your proven knowledge of the course.

5. Carefully fill in your name and address in the spaces provided and send your completed examination to:

Certification Examiner c/o A. Ruiz
36-Hour Statistics Course
Professional & Reference Division
McGraw-Hill Book Company
11 West 19th Street
New York, NY 10011

---

1. A professional survey which is designed to predict the success of a proposed new product is an example of __________.
   a. inferential statistics
   b. descriptive statistics

2. An annual report which shows quarterly earnings is an example of __________.
   a. inferential statistics
   b. descriptive statistics

3. A company takes a survey of 200 of its employees for the purpose of determining employee morale. For this study are the 200 people in the survey the sample or the population?
   a. the sample
   b. the population

4. You read a study which says that 40 percent of your customers are "satisfied" with your service. When you ask, "How accurate is this percent figure?," you are asking for __________.
   a. the margin of error of the estimate
   b. the variance of the estimate
   c. the probability of the estimate

5. A database of employee information is set up to include these variables: age, sex, years at previous job, and job category. Which of the variables are categorical?
   a. age and sex
   b. age and years at previous job
   c. sex and job category
   d. job category and years at previous job

Questions 6 to 8. Data was collected on the years of service of the employees in a company. Read the stem and leaf diagram and answer the following questions.

Years of service

Stem	Leaf
0	1 2 2 6 8
1	0 2 3 3 3 5 7
2	2 5 9
3	1 4

6. How many people were in the total survey?
   a. 4
   b. 17
   c. 34
   d. 6

7. How many people were employed at least 20 years?
   a. 12
   b. 5
   c. 34
   d. 7

8. What was the median length of employment?
   a. 17 years
   b. 15 years
   c. 13 years
   d. 10 years

Questions 9 and 10. Refer to the graph. A production run of ball bearings was produced and their diameters measured. The ball bearings were supposed to be one centimeter in diameter.

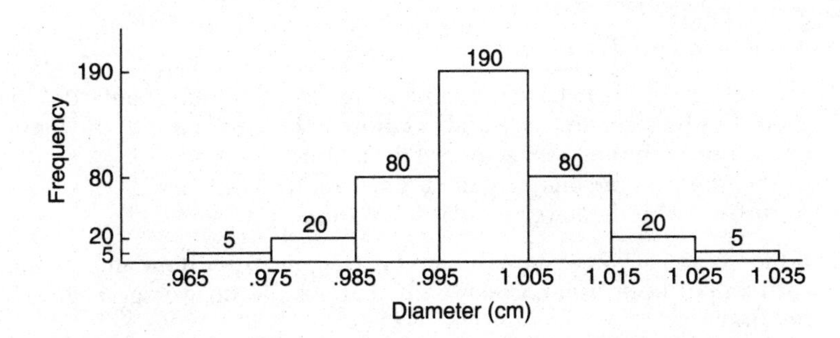

9. How many balls were tested?
   a. 190
   b. 400
   c. 100
   d. 1035
   e. 295

10. What percentage of the ball bearings were within .015 centimeter of the desired diameter?
    a. 47.5%
    b. 87.5%
    c. 42.86%
    d. 12.5%

11. The graph shows the approximate distribution of retail prices for a prescription of a particular medicine as recorded in 1000 randomly selected pharmacies across the country during the same month. From the graph determine which is higher, the mean or the median price.
    a. the mean is higher
    b. the median is higher

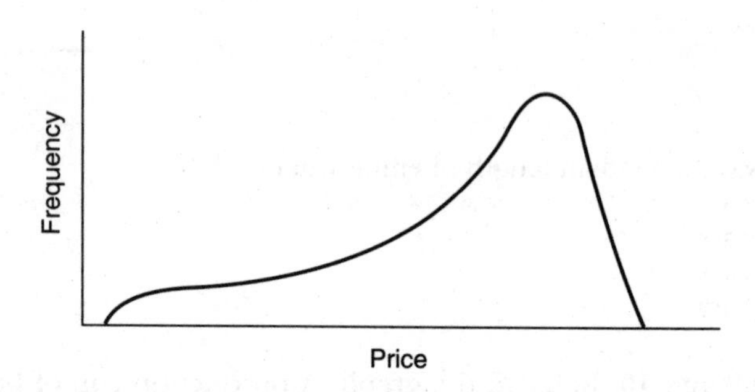

12. For which of these measures is it generally true that half the measurements in the data set are higher and half are lower?
    a. the mean
    b. the median
    c. the mode
    d. the standard deviation

13. Two sets of numbers have the same mean, but different standard deviations. In which set are the numbers more alike, the one with the higher or the one with the lower standard deviation?
    a. the one with the higher standard deviation
    b. the one with the lower standard deviation

14. If the women at your company earn on average $6 per hour and the men earn $8 per hour, is it correct to say that overall the average wage is $7 per hour?
    a. correct
    b. incorrect

15. True or false? If the mean salary for a department is $25,000, it is safe to conclude that about the same number of people earn more than $25,000 as earn less than $25,000.
    a. yes
    b. no

16. The time taken to serve customers in a fast food shop was recorded for several hundred customers. The time taken to serve one particular customer had a standardized score of –1. Was this customer served relatively quickly or relatively slowly?
    a. relatively quickly
    b. relatively slowly

17. A company manufactures computer ribbons, the lifetimes of which are approximately normally distributed. The mean lifetime is 2 million impressions, with standard deviation 100,000 impressions. About what percentage of the ribbons will fail before 1.9 million impressions?
a. 25%
b. 50%
c. 34%
d. 16%

18. A production manager is responsible for the production and packaging of tape dispensers. She notes that in a random sampling of 1000 dispensers, 7 were defective. She concludes that there is about a 0.7 percent probability that one of "her" dispensers is defective. This is __________.
a. empirical probability
b. theoretical probability
c. subjective probability

19. If the chance for success of a business venture is "1 in 10," what are the *odds* against success?
a. 9 to 1
b. 90 percent
c. 9 out of 10
d. 10 to 1

20. For a venture to be successful two independent events must both occur. There is a 40 percent probability that the first event occurs and a 50 percent probability that the second of these events occurs. What is the probability the venture will be successful?
a. 25%
b. 50%
c. 10%
d. 90%
e. 20%

Questions 21 and 22. Use the given table. The table shows the number of phone contacts made in various sales regions and whether or not sales resulted.

	Region A	Region B	Region C
Sale	20	10	40
No sale	380	190	360

21. Find the (empirical) probability that a phone contact results in a sale in Region A.
a. 20/400
b. 20/380
c. 400/1000
d. 20/70
e. 20/1000

22. True or false? P(sale given region A) = P(sale given region B)
    a. true
    b. false

23. A quality control test shows that 291 out 300 pieces are good enough to be shipped. Based on this, what is the *failure* rate per one thousand pieces made?
    a. 300 per thousand
    b. 30 per thousand
    c. 3 per thousand
    d. 9 percent
    e. 3 percent

24. A company classifies its sales accounts into large, medium, and small. To estimate the amount outstanding at the end of a quarter, an auditor selects at random 50 percent of the large accounts, 25 percent of the medium accounts, and 10 percent of the small accounts. This constitutes __________.
    a. simple random sampling
    b. stratified random sampling

25. There is a 10 percent chance that a speculation will result in a profit of $500,000, but there is a 90 percent chance it will result in a loss of $25,000. What is the expected value of this speculation?
    a. $27,500
    b. $475,000
    c. $23,750
    d. $26,250

Questions 26 and 27. Refer to Figure 4-3 on page 91.

26. How many women in the 30 to 44 age group were in the labor force as of June 1988?
    a. about 21 thousand
    b. about 21 million
    c. about 33 thousand
    d. about 33 million

27. True or false? The birth rate for women in the labor force age 30 to 44 was 33.1 percent.
    a. true
    b. false

28. For inspection, you pick one piece at random from a production run of 50 pieces in which there happen to be two defective ones. What is the probability that you pick one of the defective ones?
    a. 4%
    b. 2%
    c. 25%
    d. 40%
    e. 50%

29. You want to estimate the median income of the subscribers to a certain magazine. This desired figure is __________.

a. a statistic
b. a parameter

30. In a survey of 1500 subscribers of a certain magazine the median family income was found to be $45,000. This number is __________.
a. a statistic
b. a parameter

Questions 31 and 32. The results of a random sample of about 1000 potential buyers of your service are presented to you. It shows that 45 percent of them have heard of your company.

31. The margin of error for this survey is __________.
a. less than 1 percent
b. about 3 percent
c. more than 5 percent

32. Is it likely that more than half of the population of potential buyers have heard of your service?
a. yes
b. no

33. True or false? If you increase the size of a simple random sample, the margin of error attached to an estimate based on the sample data will decrease.
a. true
b. false

34. When a probability refers specifically to only a subgroup of a population it is called __________.
a. conditional
b. empirical
c. subjective
d. theoretical

35. If the same percentage of men and women are promoted in a company, then the variables gender and promotion are statistically __________.
a. independent
b. dependent

36. You run a placement agency. You send out a mail survey to 500 randomly chosen clients from your files. In the survey, you ask them to indicate their current salary. You get 250 returns which include the information you want. Is it reasonable to treat this data as a simple random sample from the population of all your clients?
a. yes
b. no

37. When a survey includes everyone in the population of interest it is called __________.
a. a stratified sample
b. a census
c. a nonsimple random sample
d. a confidence interval

38. The margin of error for an estimate is approximately two standard errors when the distribution is __________.
  a. normal
  b. chi-square
  c. empirical
  d. skewed

39. A 95 percent confidence interval has what probability of not including the correct value for the parameter it is intended to estimate?
  a. 50%
  b. 95%
  c. 5%
  d. 2.5%
  e. 1%

40. True or false? The standard error for an estimate of a parameter is the same as the margin of error.
  a. true
  b. false

41. In a hypothesis test for comparing two populations, the null hypothesis states that the two populations __________.
  a. differ
  b. do not differ

42. When a study compares random samples from two populations and results in finding a "statistically significant" difference between them, this means that __________.
  a. the difference is of great practical importance
  b. the difference is probably just due to the peculiarity of the samples picked at random and would not hold up under further investigation
  c. the difference would probably continue to be found if both populations were completely surveyed
  d. the finding should be published because it is a major contribution to research

43. A test is set up at the 5 percent significance level and the p-value turns out to be .03. Does this suggest that the null hypothesis is probably correct or probably incorrect?
  a. probably correct
  b. probably incorrect

44. In testing the hypothesis that four ethnic groups in a large metropolitan area have the same percentage of applicants approved for consumer loans, the appropriate test statistic has a __________ distribution.
  a. chi-square
  b. normal
  c. t

45. Which design usually has more power to detect a difference between the means of two populations?
  a. matched pairs design
  b. independent samples design

46. On the basis of data from two independent random samples you conclude that two populations have different means. It turns out later that you were wrong because just by luck you happened to get two untypical samples to compare. Which type of statistical error was committed?
    a. Type I
    b. Type II

47. When scientists compare identical twins to ascertain the influence of heredity on some outcome, they are using __________.
    a. independent samples design
    b. randomized blocks design
    c. unconditional probability
    d. stratified regression

48. A statistical technique for using values of one variable to make predictions about the value of another is called __________.
    a. regression
    b. exponential smoothing
    c. time series forecasting

49. If you use three predictor variables in a regression model you are doing __________.
    a. simple regression
    b. multiple regression
    c. nonlinear regression

50. One regression model has $r^2 = .96$ and one has $r^2 = .04$. Which one will have a smaller margin of error on its estimates?
    a. the one with $r^2 = .96$
    b. the one with $r^2 = .04$

51. True or false? The high correlation between years of education and lifetime earned income proves statistically that more education leads to more money.
    a. true
    b. false

52. Which of the three given numbers is most likely to be the correlation coefficient between the weights of delivery vans and the miles per gallon they get?
    a. $r = 0$
    b. $r = +.8$
    c. $r = -.8$

53. A report says that there is no correlation between family income and the amount of money families spend per year on video rentals. This means that in general the more money families have the less they spend on video rentals.
    a. true
    b. false

54. You want to smooth a series consisting of five years of monthly sales figures by using a moving average. How many observations go into the calculation of each average?

a. 60
b. 12
c. 39
d. 20
e. 10

55. Do you expect a seasonally adjusted series to be more or less smooth than the original series?
a. more smooth
b. less smooth

56. For this series and its smoothed version, compute the MAPE.

Time period	1	2	3	4
Original series	100	110	116	124
Smoothed series	90	105	110	120

a. 5.76
b. 25
c. 6.25
d. 177

57. In a quarterly series covering 10 years, a clear relationship exists among all the first quarter values. This is called __________.
a. autocorrelation
b. regression
c. secular trend

Questions 58 to 60. Use the given series. The numbers represent annual profits in millions of dollars.

Year:	1986	1987	1988	1989	1990
Profit:	1.5	1.8	2.5	2.7	3.3

58. If you create an index with 1986 as the reference year, what is the value of the index in 1990?
a. 180
b. 220
c. 145
d. 240

59. What is the percentage change of the index between 1988 and 1989?
a. 2%
b. 20%
c. 8%
d. 18%

60. Use the consumer price index figures in Table 9-7 on page 208 to express the value of the profit in 1990 in 1982 dollars.
a. $2.52 million
b. $1.38 million
c. $2.70 million

Name______________________________

Address______________________________

City________________State____Zip______

Final Examination Answer Sheet: The McGraw-Hill 36-Hour Statistics Course

See instructions on page 1 of the Final Examination.

1. ______	13. ______	25. ______	37. ______	49. ______
2. ______	14. ______	26. ______	38. ______	50. ______
3. ______	15. ______	27. ______	39. ______	51. ______
4. ______	16. ______	28. ______	40. ______	52. ______
5. ______	17. ______	29. ______	41. ______	53. ______
6. ______	18. ______	30. ______	42. ______	54. ______
7. ______	19. ______	31. ______	43. ______	55. ______
8. ______	20. ______	32. ______	44. ______	56. ______
9. ______	21. ______	33. ______	45. ______	57. ______
10. ______	22. ______	34. ______	46. ______	58. ______
11. ______	23. ______	35. ______	47. ______	59. ______
12. ______	24. ______	36. ______	48. ______	60. ______